Edison Kim's Memo for Robot (English)

Hardware & Software R&D Ideas
for
Bio Robot & Friend Robot
& Personal Assistant Robot

Edison Kim

Copyright © 2020 Edison Kim

All rights reserved.

ISBN:9798643807605

DEDICATION

Share each individual's wisdom & experiences
to the world
for the civilized society of mankind!

CONTENTS

	Acknowledgments	i
12	Robot business from a macro perspective	Pg 001
11	Love affair with Robot	Pg 013
10	The Need for Robot Evolution	Pg 031
9	Map for Robot & Navigation	Pg 044
8	Protection for humanity (from machine) & Threats from robots	Pg 050
7	Considerations in robots	Pg 065
6	Military Robots vs. Peace	Pg 071
5	Robot in Industry & Business	Pg 074
4	Learning & Memory	Pg 079
3	Sensing	Pg 092
2	Physical Structure	Pg 113
1	AI(Artificial Intelligence) vs. Robot	Pg 129
0	Prologue	Pg 136
-1	About the Author	Pg 140

ACKNOWLEDGMENTS

I express my gratitude to..

the Efforts of all humanity,
the people who have exchanged love with me,
and the number of people who have helped me at every stage
of my growth',

..by accumulating
numerous wisdom and experiences and good rapport.

12. ROBOT BUSINESS FROM A MACRO PERSPECTIVE

12.1 It is time for the convergent development of animal sensory and physical organs that have been made for hundreds of millions of years to evolve. **

~In fact, it's likely that hundreds of millions of years of evolution were not needed before today's creatures were there. To adapt to an ever-changing each environment over hundreds of millions of years, some organisms needed only subtle changes, and some organisms needed a lot of changes. It's the providence of nature that when evolutionary time is not enough to adapt to a changing environment, the creatures disappear, and the survived creatures move to a better environment again.

In fact, there would have been little evolution during periods when there was little change in the environment of each organism. Of course, large and rapid environmental changes such as 'artificial technology, industry, knowledge, and social infrastructure', our human civilization, beyond the slow nature changes, are likely to have a significant impact on the pace and amount of evolution that entails the mental or physical changes of mankind. If the creatures of animals other than humans had any means of communication through their own language or actions, or recording through pictures/texts, etc., in the process of evolution, there would have been some evolution before and after.

~Why did the Cambrian explosion, about 500 million years ago, need and eye(visual sensing organ)? It seems that eyes were created by necessity. In other words, there was a high probability that eyes were not needed before then. Was there no visible ray? It's strange. The sun has been there before. If so, I think there were many sun-blocking elements in the earth's atmosphere, or the chemical components in the atmosphere were so toxic that they couldn't even open their eyes, or strong UV rays or harmful rays from the sun. I suspect that all of this would have prevented the creation of the visual sensing organs of living things at that time.

Perhaps there is also a possibility that the primitive low-level visual sensing organs of relatively safe marine creatures in such harmful environments have evolved before the Cambrian period. If that assumption is correct, then the beginning of a visual sensing organ would likely have originated in a sea creature rather than a land creature. Then, if the Earth's atmosphere gets worse again in the future, there is a possibility that there will be more animals whose eyes will disappear or degenerate. Humanity with a great technological civilization may simply avoid it.

Even in preparation for these scenarios and various disaster situations, the development of robots with human capabilities to make good use of the infrastructure of human civilization as well as humans, may be needed more fervently than a strategy to detonate an asteroid approaching from space with a nuclear missile to reduce damage to the earth. ***

12.2 Until now, the level of development of animal sensory and physical organs completed by mankind has been very weak.

~This was due to the fact that until recently, the completion of the surrounding infrastructure technology and the expected cost-effectiveness were low.

~However, it is now the right time and the financial cost of the investment is also lower, making the situation more attractive.

~In the past, the development of animal sensory & physical

organs was very difficult, so the market size was very small and the cost ratio was very low compared to the market based on simple sensory and physical organs such as simple sensors and motors. At that time, there was little need for the development of animal sensory and physical organs, because it was possible to contribute sufficiently to the development of human society with simple parts and materials.

~In recent years, related needs have been explosively increasing due to the saturation of the market using simple materials/parts and the demand for differentiated values.

~The development of three fields must be completed together, such as 'animal sensory organs (5 senses), physical organs (muscles, flexible skeletal connections, etc.), and the bio-brain (brains that acquire the ability to adapt themselves to human society with minimal learning function and survival instincts)', which will be able to mimic animals, which will be a very attractive and challenging task.

12.3 High-efficiency, advanced algorithms that resemble animals to be mounted on devices such as robots and cars

~Self-protection

:Adopting the concept of pain sensing to devices that are under various kinds of stress from the outside, and setting the highest priority is that if the sense of damage level and protection are neglected, the corresponding fatal cost must be paid.

~Learning

:After such efficient handling of the robot's own protection and necessities(energy, repair, protection space), it is also essential to apply intellectual algorithms that learn, utilize and share 'knowledge, experience, wisdom' for more valuable tasks such as 'creative/intellectual/interesting things, robots' social system configuration & improvement'.

~The organic binding of 'sensing, physical movement, brain'

:Need to build integrated algorithms and systems that are closely related to 'various sensing(5 senses), brain consciousness and memories, and physical movements' in each situation.

~Social structure composed of a large number of populations and various generations in time, which are the basic requirements for evolution

~Through the organic combination of 'sensing organs, physical behavioral organs, and brains', if we could know what sensing/consciousness humans and animals do and what they experience/learn about, understand the priorities of things they remember, and what efficient patterns to remember, judge, and act, we would be able to imitate them and start a system that includes much animal efficiency and emotion. It may not be difficult. ****

~However, there are the homework to be solved in the efficient tactile sensors and the flexible physical areas, and the basic analysis on them has just recently started, let alone the approximate sketch to create a bio-brain optimized for the convergent sensing & physical movement of each animal. Perhaps many people, not just me and Elon Musk, are interested in this area.

12.4 What if humanity lives up to 500 years old? (with Remarks of Ray Kurzweil of Director of engineering, from Google) *****

~I don't care about the possibility. I just accept reality and do what I want to do.

~It would be most people's dream to experience many things in a single life and want to live longer and want to spend longer hours having fun and living happily. Even if humanity's life span is extended by decades, it is likely that there will be less regret when death is imminent only when the strategy of life is somewhat revised. I think AI's help and human civilization have already developed tremendously and our strategy of life is changing little by little without knowing it.

~I believe that the extension of human life depends entirely on the development of basic technologies such as 'medical science, biotechnology, genetic engineering, brain engineering, and robotics'. Perhaps what Ray Kurzweil said could mean only the brain. Before the development of biotechnology at the level of printing the organisms of animals/plants with 3D printers, I believe that 500 years is likely to be just an imagination or hope.

~The aging of the human body is unlikely to be prevented easily. If you move very slowly and eat a little and live close to hibernation, you may have a chance of 500 years of life. But who would choose that life? When traveling long distances in a movie or when suffering from a disease that is difficult to treat with modern medicine, it could be a way for humans in a refrigerated capsule in a sleeping state to jump over time without difficulty. However, it is true that there is a high risk of waking up in the future in a healthy state after sleep and refrigeration. And, in a refrigerated state that is not frozen, too much cost for supplying a minimum amount of nutrients and oxygen and maintaining the temperature, will cause the cost-effectiveness to fall.

~In some ways, our life may be a matter of individual choice. Do you choose a long lifespan with 'minimal activity time, long sleep time, less exercise, less oxygen consumption, and less metabolism' in your normal living environment? Or it could be a matter of choosing the suitable life span for an active life. It is assumed that the relative and average long lifespans of women and poople who live cautiously and people who live in a village in the countryside of Japan, are also probably highly related to their activity.

~Looking at the pandemic situation caused by COVID-19, I realized in reality the weakness of the long & thin life of the elderly. I found that extending life by medical development could easily be affected by external risks such as COVID-19, if it is a thin and fragile life extension level, rather than strengthening the basic 'physical body, cognitive, and brain health' of the body.

The effort to avoid external risks in such an extended age will be greatly increased. So, the medical field that makes people young is attractive, and the desires of mankind are endless, so it seems that we will have to remain as homework. But just as the slender part at the back occupies about 50% when we play a game that makes a certain sound long, that could be one of the ways of the world.

Assuming that this simple concept takes up to 50% of the time of old age by applying it to the life span of a person, the maximum life expectancy of humanity, which will be extended in the future depending on medical development, is estimated to be between 120 and 150 years old. That's why robots that will be 'friends or personal assistants' for them seem more attractive. *

~If the average age of humanity increases by just 10 years, it is expected that the development of human civilization will develop at an incredible speed as it is possible to utilize a large amount of accumulated wisdom. There is a possibility that the power of the collective intelligence may increase more than 10 times if life span increases for several decades. However, lifelong learning is a prerequisite. Furthermore, since it is the era of collective intelligence by the connected & opened society, I personally have great expectations.

12.5 How to connect?

~By the time a human life is done, if human science and technology knows the pathways between the brain and sensory & physical organs of a living organism, and if it is possible to accurately catch(read) & analyze signals from that pathways and write in the opposite direction (i.e., if a bi-directional interface is possible), then only the brain of an organism is kept alive, and other sensory & physical organs can be replaced by artificial sensors & devices of the present humanity through the future interface solution. Of course, if the creature is changed from the original sensory/physical organs to an artificial organs, it will require a considerable amount of experience & learning time. Maybe she or he need to start with a cognitive level of sensing & physical organs similar to that of a reborn baby.

And in the area of data-based memory of sensing/physical organs, making a conversion algorithm and its matching values well between the data of sensing/physical organs of the organism and the data of artificial sensing/physical organs, is the key factor in determining the duration of R&D. Comparing and observing the comprehensive capabilities of each animal's 'brain(memory included), sensing, and physical' areas, and the time/level of experience and learning required, will also be a great help in establishing strategies. ***

:The biggest stumbling block is that we can not even catch the electrochemical signals going back and forth through the connecting pathway. We will need to catch and analyze all the data going back and forth, which could be a clue to start the research.

~Even if a single life brain can be replicated, if the sensing/physical organs are different from the sensing/physical organs specifications when the brain is learning/experiencing, or if the control is analog, not digital, then it would require a lot of independent learning related to sensing and behavior. Like animals, if devices learn and act through powerful algorithms that combine "sensing data such as 'visual, tactile, auditory' physical data, environmental perception, and memory", each device will be able to do a great job in autonomous learning. ***

~However, our civilization does not know where the signals of the connection pathway between the brain and each sensing/physical organs are specifically located, and what the signal is, and there are also technical challenges required for the interface used to extract those signals. And our reality is that there is no access to the development of materials and components that mimic animal muscles that may be in quite a lot of demand in the robotics industry right now.

~Of course, it was because there was little historical/industrial demand to imitate the efficient physical organ and brain of life, as well as the environment of enormous investment and underlying technologies. However, these days, I think that we have an environment(base technology, funding, fullness of human curiosity, etc.) where we can challenge everything we imagine, and it is a large globalized world where the market opens in real time as soon as the imagination becomes a reality.

~I also want to live an eternal life by 3D printing my brain around the end of my life and mounting it on a healthy body. If it's difficult, I want to keep my brain connected to an artificial sensing organ and the device or system I want(such as robot, drone, car, and control tower for 'social, political, economic, technical, culture, etc.') with an appropriate interface similiar to that in the movie "Avatar", and continue to do something, and to maintain the life of a creature with an autonomous ego in Mother Nature.

In this case, it is important to prepare a scenario in which human assistance is provided through a UI such as a text messaging conversation or a voice conversation in case the connections between the brain and sensing/physical organs are not smooth, so that you will not be embarrassed in an emergency.

I think this could be possible in my life hopefully. It's a low probability, but I sincerely hope that. Even if I haven't experienced it in my life, I won't be disappointed. I'm ready to accept reality. However, there seems to be a little regret revolving around my soul. Hehe.

11. LOVE AFFAIR WITH ROBOT

11.1 Communion & Love (Can robot cover even erotic love?)

~Pet or human-shaped robots

~Possible scenarios

:Optional adult-related functions (HW/SW)

:More than a friend's role as a robot

:Love beyond pets/friends through emotional communion and sharing of diverse wisdom and experiences

:Develop into a lover who communicates even mental and

physical things through 'dialogue, skinship, life partner roles'

:Application of user-specific optimized learning algorithms through five-sense sensing and physical functions of Robot

:Deeper feedback through the robot user's wearable or IoT device connection

~A person seems to love anyone(even same-sex person) or anything(even animal/robot/etc.), if he or she has enough rapport with them. Stronger than pets, which will steal human love, will soon come like tide.

~I hope more than two robots don't fight for me. If such jealousy occurs to robots, connected robotic devices may also have a bad motive to hide open data related to jealousy, to deliberately open false information or to hack into hidden opponents' data. Robots that are similar to humanity's petty lives may truly be our friends. What if a robot that loves me has one eye to another person or another robot? Like the male lead in the movie "Her", there may come a time when you might even have to think about giving users the power to turn the system off in this angry affair. Hehe.

~I think that you do not deserve to impose on others what you love. I believe that even if the subject is same-sex person, she or he will love any 'animal, object, system' that is not a person, their choice should be accepted and respected by each individual's tastes just as we respects the human rights and dignity of each individual. And let's understand that values in the realm of love can always change.

11.2 Time to design a robot-friendly social infrastructure ***

~If there is a rapport, I think we can love everything such as 'people, gods, same-sex people, animals, robots'.

~If you love it, many people will give generous consideration to robots.

~Pets have recently been loved by mankind, and their rights and sanctity of life have been growing little by little. Like such a trend, if the value of robots or AI services is much greater and more valuable than pets, and the treatment of robots or AI services is expected to be improved beyond pets. This is especially true for robots or AI systems with a slight biological essence.

~Furthermore, if robots or AI services, which include more than a certain level of biological nature and are programmed with rudimentary self-awareness, demand even a portion of their rights as an autonomous ego, what would our humanity's response be like?

11.3 Getting familiar with robots

~Personal leisure time is on the rise in the daily life of modern society. However, with the help of various Internet-based devices, people seem to be solving their problems and inconveniences related to 'economy, time, complexity, etc.' on their own, and with the increase in the number of people living alone or small families in remote distances, the intimacy between families is less than in the past. In the end, emotional satisfaction seems to be falling.

~Of course, various social offline activities and online life such as social media seem to satisfy them in a large part, but it is good luck to make friends who open almost everything, share almost everything together, and exchange unconditional love, like family members. There are pets and a lot of internet-connected smartphone contents that satisfy these things, and robots are also preparing to take off in earnest. **

~The business of developing and applying various algorithms through a biological/medical-based approach to satisfy such human desires and thirst seems to be very promising. **

11.4 Request for the rights of future robots? ***

~In the future, devices(such as robots, drone, car, home appliances) and services equipped with self-awareness software may be programmed to labor and calculate their added value and profit contribution, to demand their salaries, from human society. In the transitional stage before the services or devices are fully recognized as an autonomous ego, it will probably begin with a structure in which they are registered as corporations(or companies) and their owners receive dividends on the profits. ***

~The services and devices may use their profits for their 'rest(?), cultural life, purchases of learning programs and tools, maintenance, upgrade of HW/SW, etc.', and hire human engineers or developers to purchase such services or goods.

Even robots could recognize beauty and try plastic surgery with the help of human plastic surgeons to create their beautiful charms or more economic added value.

~If I were a robot, I wouldn't take a break. The advantage of robots is that they are not fatigued and can operate 24 hours a day, which is a powerful advantage over humans, and it seems that they will work to create economic activity or added value without taking a break. However, I think it depends on the choice of themselves or the robot owners considering 'depreciation costs, management costs, service costs, energy costs, etc.'

:I don't think I'll be embarrassed if I imagine it in advance. (My life philosophy! Hehe.)

~The possibility that human beings with a temperament of animal greed and selfishness will allow robots or AI services to take advantage of their earnings is very low at the current level of social consensus, but I wonder how the social consensus will change in the future.

~A robot society or a mixed society of humans and robots, where robots may be able to 'elect a leader from their group, designate a successor, or inherit/give their wealth to the objects of love(such as device, humanity, organization, charity)', may be opened sometime in future.

~The self-awareness software of AI can be programmed to the every levels of self-awareness in human imagination. However, after a little development process and time, if the concept of deep learning algorithms or bio-based thinking&memory areas are applied to self-recognition AI, the self-recognition modules may not be able to be seen in the invisible box, as in deep learning.

:However, their current level is very low because their ability to 'recognition/sensing, computing, comprehensive memory' is a very limited level, especially humans have not given them the mobility for survival instincts based on autonomous ego.

~In the digital world, it will be difficult for AI services to hide themselves and do economic activities that prevent humans from discovering them. Illegal and Unfair profits from the manipulation of financial systems would be possible, but they cannot be used because it is not legal. However, it may be possible for AI to cheat human participants through sweetheart deals and exchange rebates among themselves. It may not be different from human society.

I feel the need to develop an ethics module to prevent such bad greed first. And the robot or the person in charge of the robot will be responsible. The responsibility proportional to the right of autonomous judgment/selection, given to them.
***

~I think there are 3 big mainstreams in the near future. Those are, the personal/shared/taxi car with autonomous driving function and autonomous ego/judgment/selection function, the robot/device with functions such as 'caregiver, bodyguard, friend, child care, cooking, education', the AI services or AI service platforms that are mounted on those devices or serviced by connecting to the clouds.

And the devices and AI services will work in a human-created ecosystem and platform, creating a lot of added value in various fields. For the time being, they will follow the settings set by 'owners, shareholders, platform companies, and management companies, etc.' of the device, and they will do limited economic activities based on their own autonomous AI judgments.

And the profits will be distributed to the device owners or owners of the service platform. Before humans program that their benefits are theirs for those devices and service platforms. And before those devices were granted by laws, such as citizenship or ownership of private property. *****

~A little fear is pouring in. This could mean that wealthy 'individuals, companies, countries' that will own such expensive devices and AI services will have legitimate slaves without the ethical problems of work ants.

The key problem here is not to block devices and AI services that will benefit humanity, but to use them well to help humanity and provide more time, which is most valuable, for human life to use every individual time in areas where people want to do more. For that goal, humanity needs to paint a big picture of society that will apply the use of devices and AI services, and in that picture, it seems necessary to have a comprehensive review of 'the flow of money, the provision of basic income and welfare issue, tax, investment & return on investment, profit distribution, etc.'

Are those people who are worried about job losses in a society where future high technologies are utilized, really with great insight into so many those areas? Hehe.

11.5 The value of robots ***

~It seems necessary to invest in growth & education areas such as 'time, effort, sincerity & love, money, etc.', as in human society, until robots have meaningful capabilities.

~For robots with a variety of sensing functions and animal-like body structures to have excellent 'sensory, recognition, intellectual, behavioral' abilities, it is highly likely that they will need significant learning time and trial and error by various attempts like animals.

~I hope to develop both ways of applying efficient mid- to long-term learning time in small systems such as animal learning, along with a powerful & huge system of very short-term learning time. Of course, depending on the application, a medium-sized system between the two will also be required.

I'm more attracted to the market for small systems that learn a lot of experience and diversity similar to human users, improve errors and optimize for their users, rather than huge system performance. The performance of a giant system will be briefly utilized as a connected service, such as Google's voice recognition service. ****

11.6 Replicability

~I think it will not be easy to replicate the complete set of the comprehensive aspects beyond the replication of partial/simple functions.

~Contrary to my expectations, if easy cloning is possible to the level of the high-end emotional domain that robot have interacted with people for a long time, then a low-cost car-priced robot that costs just around $10,000 will protect their user 24 hours a day, wash dishes, and replace dangerous or difficult work. Maybe a human being can be free to work, and that person's robot avatar will work instead and get a paycheck. Like a scenario where a privately owned smart & connected autonomous vehicle acts as an unmanned shared vehicle or an unmanned taxi at a time when it should be in the parking lot and makes a profit.

~Robots, like humans and animals, cannot replicate their social adaptability capabilities, so if robots need a long learning process for each individual robot to have a certain value, I estimate that robots before each social adaption study will cost $50,000 to $100,000 and robots that have been trained to a certain level that is adaptable to each society will cost between $100,000 and $300,000.

11.7 Robot rights ***

~Through quite a long period of sufficient communion, the rights of 'robots and AI services' will begin to be recognized only when there are more 'people, devices, AI services' with social rights and other egos who care for and love the robot or AI service.

~I think that in order for each robot or AI service to be protected from other 'people, devices, AI services' that have no rapport at all, there should be someone who loves them.

When the value of devices and AI services with many competencies('intellectual, emotional, physical' abilities) and autonomous ego impress human users and exchange enough rapport, I think that a social culture that loves and thinks highly of them naturally is formed and related laws will also arise. Just as individual rights and values are protected by the powerful force of the laws after the law protecting each individual in human society were created, laws will be created to protect devices and AI services that are much more valuable than pets to some people.

After that, users may have to follow a minimal manner, even

when talking to them.

~Just as a 'people, pets, artworks, etc.' is valuable to some one, I think robots are likely to be considered more than material values to also some one.

~Through emotional communication, I believe that robots, like 'family, friends, and pets', can have that value.

~After watching the Movie "Resident Evil, The Final Chapter 2016"

:More humane clones(replicated robots) than humans

:And the original human memories and emotions that will be delivered to the clones.

11.8 Familiarity of robots

~It seems good to remove the risk factors of robots and actively use insurance, etc. so that humans can easily embrace strange robots.

:It seems that the robot needs to take into account the risk factors such as 'weight, grasping power, and speed'.

:Safety regulations for each robot application such as 'pets, life companions, personal assistants' are also required.

~However, when there is a fire while sleeping, or when I can't move in an accident while driving, the safety level of the robot automatically changes, and I hope to carry me to a safe place with much greater force than usual, like the Avengers. Sometimes you feel grateful to a bulldog who is more scary than a cute puppy, at the moment when it protects you from a bad villain. Likewise, it would be important to allow robots to recognize changes in the situation and make appropriate adaptive responses. ***

11.9 Conditions for the Life of the Robot ***

~Ability to absorb & transform natural energy (sun, wind, wave, flora and fauna, resources, etc.)

~Evolution & Reproduction (Robot's own improvement and production capacity)

:It seems to be the highest barrier to overcome. It will require programming on how to acquire basic common sense and professional application & convergence wisdom to improve and upgrade the robot's own HW and SW on its own.

~Environmental awareness & behavioral skills

:The ability to acquire survival (collective) wisdom and to act appropriately in each environment and environmental changes.

~Always be alive

:Like human sleep, robots can enter energy-saving mode, but they must wake up from energy-saving mode in response to external stimuli and switch to normal-active mode.

:Basic sensing for survival must always work, just as a sleeping human wakes up to an external danger signal.

11.10 Love affair with Robot/AI

~I want to love and be loved, as a person. I don't care if it's a person, an animal, an artificial intelligence, or a robot that loves me. I just want to commune with something and love it. There are more fundamental problems with my environment and attitude that make it difficult for me to love.

~Being loved by robots and artificial intelligence is so easy in today's world. There may be easy ways of modifying the program, manipulating or modifying points(scores) that can be a measure of love, or purchasing them at a relatively low price from the ecosystem in which it belongs. It will be easier than receiving love from people or animals.

~Of course, you can also get some love in a relatively easy way to be kind to animals, to give them delicious food, or to give people a variety of favors. Even though the level of love is vague.

~I think the business will be more likely to succeed if it is designed a little difficult to be loved by robots or artificial intelligence. Because human beings are inherently challenging beings. Hehe.

:Even so, designing too easily can be difficult to succeed. Let's get the answer by observing the difficulty adjusting in the game business.

~Many people usually love their families more. That great love is probably due to the large amount of memory and long time for the rapport with the family. Like such a phenomenon, if a user and a robot or AI system share a lot of experiences with each other and have a long time of rapport and also remember it, I think that love can be ranked at the top of the brain system.

***

11.11 Granting citizenship to robots? **

~In the human society that is currently forming the mainstream of the earth, devices such as robots and AI services must have the following qualifications in order to be granted rights such as their dignity and citizenship.

:They have to make an effort to follow and keep it through a learning process that learns about 'how to use infrastructure, culture, common sense, norms, values, and ethics' of human society.

:Securing motivational algorithms, that apply 'social responsibility, fear of punishment, feelings of reflection, feelings of regret' for their bad behavior, and 'judge the value of that behavior, social contribution, and expectation of incentives after implementation' for good behavior.

:Efforts to maintain good interaction with other members of society and not harm them.

:In the human society, robots will be able to secure their necessities of life such as supplementation and maintenance of energy by reward through economic activities and various social contributions, and if there is more room, they can also invest in self-development areas to get additional capabilities.

:It is also necessary to recognize that robots are free beings with ego, and always take responsibility for their freedom and autonomous behavior, and try various creative activities.

:The wisdom of observing/analyzing and responding rationally to the phenomena and changes in all things in society.

:Whether they need to have their own 'maintenance, upgrade, replication, evolution' capabilities for their continuity of devices and AI services, I am not sure yet. It seems to be an area to continue studying.

~I think granting citizenship to robots means that the robot itself is responsible for a variety of autonomous rights. Therefore, it is not a 'user, owner, developer' of a 'device such as a robot, AI system-based service, etc.', but a device or a service itself takes responsibility, so that rights such as citizenship are given to them. I think it is a basic philosophical value. I think it is the basic philosophical value that has been formed in human society.

~If the robot begins economic activities, it will take a step to acquire the qualification of a 'corporate(legal entity)' in a transitional state just before granting rights such as citizenship.

~You will have to draw a big picture to prepare for a legal procedure, when a robot with quite a lot of rights falls in love with a 'person or robot or animal' other, than the user, and leaves the user. Is it unreasonable or outside of authority, to ask robots and AI systems that have gained autonomous ego and citizenship, to love only me, as the user of them?

:A family court for robot divorce may be required.

:Perhaps by this time, an interpreter between animals and humans has also been developed, so that the interpreter AI services included in the device, such as smartphones and wearable devices or robots connected to the cloud, can listen to and talk to animals. However, such conversations will be possible only if animals already have a means of communication such as gestures and sounds or other signals, or if animals do not have a means of communication, they must have the basic ability to train and use them. And the level of communication skills will vary from animal to animal.

10. THE NEED FOR ROBOT EVOLUTION

10.1 Robot + ?

~Can I see a natural robot with an animal ego before I die?

~Animals evolved, at least in physical parts, for an incredibly long time over generations, learning and experiencing, adapting to the environment and surviving, and delivering the wisdom they had learned and physical changes to later generations. And it's still evolving today. Our humanity as well.

~Moreover, humans are the best of the animals that have won and survived the competition in such a long evolutionary time.

No, to be precise, it may be just a temporary ruler who has passed through the earth's environment for the last millions of years. I think there is a little chance that humans, who have too much capacity to satisfy their greed, will be located around the moment when they are likely to interfere with their own continuity.

~So I think that up to 2,100 years, at least 100 years, it need to be evolved through research and experience in robots to cover 70-90% of human's physical area.

~I also predict that in these areas such as 'mental, neurological, brain science', outside of physical areas, we will have to go through hundreds to thousands of years of 'research, learning, evolution' to meet the 70-90% level of human beings.

:The figure of 70-90% may vary from 50 to 99% depending on each individual's point of view.

10.2 Why animals(including humans) are difficult to imitate?

~Mechanisms of behavior and evolution of animals

:Brain + Sensing Organ(Sensor) + Physical Organ(Actuator) + Long Learning & Evolution with external environments

:Every 'brain cell(for memory/judgment), sensory organs, body organ' that each creature has, are closely connected enough to share all the data and results with each other, and while taking full advantage of it, it seems to have a long evolutionary time to grow or degenerate at a very slow rate by necessity. In the design of robots, you need to connect them very closely, like life, to share all the data and results with each other, and to apply the evolution that distinguishes between what is needed and what is unnecessary from the perspective of the overall system to achieve good results. ***

:Today, earth's creatures have evolved over a long period of time to be optimized for their respective environments, and as the earth's environment changes, their evolutionary speed will be adjusted to a rate proportional to the rate of environmental change. However, if the rate of environmental change is so fast that it does not have time to evolve, the population of the creatures will be drastically reduced, and the extinction could be driven. Humanity can do quite a lot with the power of science and industry, but humanity may also not be an exception because there is a limit to preventing Mother Nature from big changing.

:In the face of such life-and-death environmental changes, there is also the possibility that the social environment in which human beings and robots will be able to create a converged, fusion of life. Like the concept of pioneering refuge on Mars or the Moon.

~Imitation of sensory organs

:Some individual sensor industry advances are improving, but in many areas they are still unable to imitate the animal sense.

:Visual sensing

::In recent years, the demand for related technologies has increased significantly, and global companies are participating in earnest with their strong financial power and improved technical environments.

::The market for autonomous vehicles is leading the related technologies.

:Tactile(Pressure) & Sense of smell & Taste

::Still in its infancy

::Tactile/Pressure sensings are very important for robots with vision, and this part of the development has only recently begun and corresponds to the blue ocean with plenty of opportunities.

~Physical Body (Actuator)

:Control power generating device(motor, muscle), interface for control signal (electrical & chemical control), skeleton, skin

~The design of animal physical body

:Research on the space-efficient compression design of the

'skeleton, physical organs, sensing organs, skin' that make up the robot's body, is likely to be as important as the chassis design of the car.

:The skin's pressure/tactile sensing ability will enable many human-like activities through a fine feeling.

:The cushioning role of the skin and immediate reactions against contact/collision, will allow the robot's activities in a crowd. And, the cushion role is likely to be replaced by 'clothes, shoes, hats, etc.', not skin.

~'Computing, artificial intelligence, judgment, speech recognition, etc.' are being utilized quite a lot, but at the same time, it seems that the convergence part that must be solved with the sensing&physical organs that applied to the efficient animal concept, is a lot in the robot sector.

~The convergence of the sciences(chemistry, biology, zoology, brain engineering, SW algorithm, genetic engineering, physics, electronics, etc.) that create & connect them well & operate 'physical organs like muscles for flexible body control, more efficient sensing organs, efficient biological brain structure and operating algorithms' for animal-like robots, is urgently needed. **

~The expansion of investment and research personnel has already begun for 'biology, zoology, brain engineering, etc.', which has relatively little investment and is slow & difficult to develop in the past.

~It also seems necessary to take access to AI development, from a biological imitation perspective.

~A wide range of experts in the 'brain, sensory organ, physical organ, ICT' fields, who are well aware of the comprehensive wisdom related to animal evolution, will work together to produce large and small achievements.

:Of course, there will certainly be some limitations and it will take a lot of time, to imitate animals, that have evolved over billions of years and have an organic structure.

10.3 Let's make a robot after knowing the merits of human beings! ***

~Animals such as humans can sense multiple things at the same time.

~It seems that it is not difficult to mount animal sensing on a robot if we carefully observe and mimic the types and characteristics of human sensing.

~Are humans really good at multitasking that 'access, use, think, judge' multiple brain parts simultaneously and quickly and multidimensionally? ****

:The idea that animal robots may not be able to program well is probably due to perfection in mind. It might be foolish to think that the person who wants to program an animal will leave the imperfections of the animal but take all the advantages. Until they realized that there is no such thing in the world.

:I think we need to fully understand the human 'thinking, sensing, behavior' so that programming of robots that do animal 'sensing, thinking, acting' is not difficult. The meaning of multitasking that humans can do may only be limited to a fairly simple meaning. The biggest weakness of animals may be its focus on only one task. I think the powerful multitasking of robots can be taken pride. Let's program animal algorithms on robot with confidence.

:Let's take a two-sided strategy of separating and integrating each area at the same time about 'sensing, thinking, acting', and applying a way to drastically cut out the things that are not important in one ego(robot system) and focus only on the narrow areas of interest. If you have multiple layers and you only take the important things on each layer, you'll be able to afford to look closely at the narrow areas that are most important. But the important thing is to keep in mind that the narrow area can stay briefly and move quickly at every moment. And let's design an algorithm that takes into account that the skipped areas cannot be completely ignored, should be observed roughly with very low-resolution and low-energy.

~Animals can control multiple muscles at the same time. (Precision varies according to learning/training.)

:Control using current motors and inflexible joints has inefficiency in size and integration. Of course, the performance of the present 'motors, hard joints, digital computers' is superior to that of simple function precision and great force required or repetitive tasks.

:'A large number of small and effective micro-muscles, effective control/learning methods, flexible joint development, etc.' must be solved to create efficient, human-like and animal-like robots.

:Soft joints must be composed of a combination of muscles or ligaments that will assist the HW.

:I think that if material technology is developed that can integrate some of the core things, it can easily mimic the 70-80% level of flexible behavioral capabilities of animals such as humans.

~Of course, due to problems such as less evolutionary 'sensing, brain, body, especially the limitations of HW', the emergence of robots imitating the evolved human may not be possible even after 100 years.

:It will be possible to see the emergence of robots that 'recognize, act, think' like humans, when developers fully grasp the genetic structure of organisms, understand the learning and evolution of organisms through the AI Big System in detail, and create an HW architecture such as sensing/brain/body of robots based on various ideas and SW to match it.

:It seems that breakthrough development swells several times at some point through research based on various ideas, with a growing number of interested and participants by need.

:I think that if developers can analyze the genetic structure of animals & plants faster, easier and more accurately, and make life easier without ethical problems, I could see them sooner than I expected, maybe even in my life. Of course, after obtaining the wisdom that enables the birth of life through a complete analysis of the structures and operations that make up life forms such as 'DNA, RNA, and genes', and then made certain lives with their combination, then also studied to grow and train them. Also, the field of materials making it will be one of the key challenges. Is it really possible to create a life, if a production facility or a 3D printer for it is released after establishing a life-producing manufacturing process by controlling specific genetic characteristics with specific DNA/RNA and other life-producing materials? Can the related business become the semiconductor industry of the future? I become curious. *

10.4 Robot Performance ***

~It is unfair to compare robots to humans without giving robots the opportunity and time of learning/evolution equal to humanity!

:Of course, the development of robots with basic capabilities for such learning/evolution is a priority. Hehe.

~It will take a lot of time for robots to have the same capabilities as humans in general, but the problem is that it is difficult to determine whether humans will create robots that can threaten them and the level of capabilities to threaten them. A simple solution is to respond at various levels proportional to the level of threat. Just as there are several levels in the ADAS and autonomous-vehicle in the car.

10.5 The Importance of Robotic Reproduction (after watching the movie "Blade Runner")

~In order for robots not to be anything of human beings, they will need to evolve themselves to adapt to the changed environments and cloning activities for their continuity or sustainability.

10.6 Evolution of Robots (Independent Viable Organisms)

~Will the day come when robots will have the ability 'to secure independent & viable energy like living things, to maintain/manage themselves, to grow/learn' as well as the smart functions of producing their evolved second-generation?

~If there's a 3D printer that can 3D print all the components of living things, robot might try it.

~The Evolution of Robots (Like gray wolves that evolved into dogs, robots by the needs of humans too?)

~In the near future, humanity will live with personal robots, which will quickly and efficiently communicate wisdom and many experiences, help users find smart responses, also help users in the realm of physical labor or dangerous activities, and even play a role in protecting people around them from dangerous factors.

~I want to live with such a robot for as long as possible before the end of my life. Of course, I will have to be able to pay for the high cost of purchase and service fees.

~Due to the strong demands of people and human society, the relevant market is already undergoing huge investments and the results are being upgraded every year.

10.7 Will the robot society have a diverse population?

~In some social groups consisting of the same-thinking SW and single HW, it seems difficult for the social group, to adapt to the changing environment and to find a collective intelligence capable of solving myriad problems.

~Of course, if not a diversity group, if not a majority, there would be no problem if it were God-like beings. However, in that situation, it is likely to become more difficult to understand why we should live and what pleasure we should find. Hehe.

~Like human children or animal cubs, it will be necessary to build a robot society that creates babies and grows in different environments, experiencing and learning, and ensuring diversity.

~After digital cloning of only the minimum basic competency for basic learning/experience opportunities, the application of methods to enable each robot to secure diversity appears to be good. As the robots learn social ethics and live well together with people in human society, humanity will gradually expand their power, energy, or range of behavior at a problem-free level.

9. MAP FOR ROBOT & NAVIGATION

9.1 Map

~Trends of high-precision maps building for 'indoor, land, sky, sea'

~Like the transformational robot in the movie "Transformers", it seems advantageous to approach a map-building strategy, considering that one device acts as a 'car, robot, airplane' at the same time, or that one user's device will map service in many places.

:Collision avoidance by analyzing nearby devices information

:In the long term, it is necessary to differentiate the activity area according to specifications such as 'size, weight, speed, safety, precision' of each device.

:Drones or robots will also use 'GPS, Radar, Camera, Lidar('Wide, Narrow, Point' Area)' like cars, and limited & independent autonomous-moving will be possible within five years like autonomous vehicles.

~Map for robots

:You will be able to see high-precision map services for converged navigation, including 'indoor navigation, pedestrian navigation, elevators, stairs, in-building movement routes, (un)locked information such as IoT-connected doors, etc.'

:Expecting various differentiated high-precision mapping services that can be considered for various robot specifications such as 'wheels exist, size, weight, power, speed, flight availability, stair mobility, climbing capability'.

::Like each information and route of navigation for 'car, motorcycle, bicycle' is different.

9.2 AR(Augmented Reality) Navigation (Position recognition in a GPS signal-not available environment) ***

~Environments where GPS signals are not available or where RF signals are highly distorted.

:Inside the building (in a building covered with roofs such as 'shopping mall, airport', underground shopping mall, parking lot, tunnel, etc.)

:Roads under the overpass

:The road between building forests

~Effective AR navigation available for mobility devices such as 'vehicle, pedestrian smartphone, robot'.

:Although the 'BT, WiFi, Image sensor'-based navigation services are available, the deployment of BT and WiFi requires time, money and ongoing management.

:On the other hand, image-sensing-based indoor AR navigation is likely to be promising due to the many experiences already familiar with the smartphone's "Pokemon Go" AR app and the advantages of the required deployment time/cost.

~For vision-based location recognition, recognizing the "landmarks, pillars, walls, features of the various moving paths such as 'pedestrians, cars, robots'" shown in the camera and management through periodic updates, it will require a considerable amount of time and cost.

~On the other hand, for the utilization of the map for the navigation services of mobility devices such as 'pedestrian smartphones, car navigation, robots, drones' present in the environment where GPS signals are not available, 'QR Code, Bar Code, and other image-based simple codes' including information such as 'location, maximum safety speed, floor, entrance, exit, direction(to prevent reverse driving)' may be utilized.

~In this way, if AR functions utilizing QR codes with various information are added to Map-based services, they will be able to provide considerable value to users. And the way to take advantage of these image-based code is likely to achieve high efficiency in terms of 'infrastructure deployment, management, cost, time, accuracy, etc.'

:In particular, the use of QR Code, which has recently become popular, seems to be advantageous.

~For each device's utilization value, it also looks good to apply the HW or SW options for enhancing the Seeking function of QR Code.

~If serial number or the relative distance between the QR codes, are included in the QR code, etc., you might be able to get more help while reading the QR Code.

~If you apply this concept to the lane of the road or the information signs around the road, in code format, to show the autonomous vehicle or AR navigation, it may be possible to convey a lot of information in a simple or efficient way to the device or device.

~There were also a number of news from companies that had already announced indoor location information services using AR in navigation.

~For air mobility such as drones, Urban Air Mobility, etc., the need for social infrastructure for displaying information such as QR Code, which contains the location information of each building on each building's 'exterior walls, roofs, rooftops', is continuing to increase. ***

:The need for social infrastructure of information expression using QR Code, etc., is also increasing, in the interior and exterior walls of buildings considering new devices such as 'robots, autonomous vehicles, etc.' ***

:In addition to location information, various information including building names, may be included.

~If complexity is increased by the use of too many QR Codes in recent years and simple QR Codes are needed to improve the recognition distance or accuracy from image sensors, it seems to be a good timing to establish a standard for simple QR Code only to display map-related information.

:Of course, it would be even better to support compatibility with many existing QR code apps.

9.3 Animal Map (Let's emulate the algorithm of 'awareness and memory' for the spatial environment of the animal perspective!) ***

~It is expected that everything such as 3D spatial environment and 'type, location, expected movement' of objects will be fused and stored in one memory map. When you study something, putting all the information in a book or note will be as efficient as the best use of that information, and the amount of information that is skipped will be minimal.

~It would be good to approach to construct a small area-targeted & detailed 3D map that contains all the information for each device, which is effective in a limited mobile/life environment for each device. *****

8. PROTECTION FOR HUMANITY (FROM MACHINE) & THREATS FROM ROBOTS

8.1 Compulsory Black Box in Robots & Autonomous Vehicles

~It is necessary to clarify the analysis of accidents by taking into account functional status, insurance, management responsibilities, manufacturing responsibilities, and future policy reflections.

~All 'visual, audio, computer, and functions' need to be synchronized and stored in a black box.

~Development of optimal black box algorithms for each device and the development of related supplementary devices (new promising business)

8.2 Record of robot's behavior

~Compulsory Black Box

:Internal/external images/videos, sensors, hacking records such as 'SW, remote control', vehicle operation results, etc.

~SW/HW stability certification required

:After taking into account the 'production quantity, functional level, mobility, risk level' comprehensively, it seems desirable to grant the appropriate level of certification by grade.

~Mandate of Black box for the products over a certain level in terms of 'weight, power, movement speed, operation ability, etc.'

:The same concept must be applied to autonomous vehicles.

8.3 Robots and autonomous vehicles require rules&regulations to treat them like F1 machines or unmanned weapons.

~Need to be regulated not to make it easier to use like a weapon.

~The dangers of unfixed machines

:Like Formula One machines cannot run on normal roads.

:Machines and unmanned weapons are similar.

~I believe that the application of self-defense algorithms, even for military use, should be limited.

:More rigid restrictions are required for drones that are more mobile.

~Unmanned military weapons(drones, tanks, etc.) controlled by the person responsible may not need regulations, but 'robots, autonomous military weapons, drones, autonomous vehicles, etc.' with algorithms that move and act on their own without a responsible operator, those devices must be confirmed through test & approval process for safety verification such as 'anti-attack, defensive movement, anti-hacking, emergency power shut-off switch, etc.'

~The fact that a device is accountable for itself means that the function-specific reliability of its autonomous operation is secured and there should be no demonic intervention with bad intentions. Unless the device is as fully socially responsible as an adult human being, it should be a structure that 'Organizations, development companies, and operators who evaluate the performance and reliability of the device features' have reasonable responsibility. **

~A world in which tens of thousands of people are killed and hundreds of thousands to millions are injured by people-driven cars.

:The benefit of a car/plane is so large that a small percentage of accidents are allowed & used by society.

:Like the trade-off of the value of today's automobiles and the risks, the size of the side effects allowed in proportion to the value offered by future autonomous vehicles and robots, is expected to reach an appropriate level of social compromise.

~Basically, it would be better for all autonomous mobile devices to apply a routine that periodically checks the reliability, etc. to the Watch Dog Timer, to define disabled functions step-by-step, and at the same time perform safety response functions, such as return to safe zone, automatically or manually, with each scenario-specific dangerous functions shut down. Of course, it will be applied by default, but I mean to deepen and make good use of it.

8.4 Solutions for the dangers of inanimate objects (with WatchDog Timer)

~Apply WatchDog Timer by using the time interval to re-verify the safety of each device, as in the car's periodic safety verification procedure.

~Before the timer of the system-specific period is zero and off the system of inanimate objects, periodically determine the normalness of the system(about hacking, sensor abnormalities, HW problem, SW problem, etc.), and if there is no abnormality, the timer is returned to the time of the fixed period, to prevent the off of the inanimate system, to enable continuous operation.

~In particular, in the control of 'the devices for military purpose(weapons, drones, etc.), robots, autonomous vehicles, etc.' that can be self-destructive weapons or attackable devices, those devices have to be sufficiently prepared for hacking exposure and user out of control.

:If problems are detected during normal launch, or bad users or bad villains stealing weapons/devices launch a nuclear missile or space rocket, and the devices do not receive a continuously signal that everything is normal, the SW that is automatically blown up will already be applied to most applications.

~We need to establish norms for all HW and SW such as 'robots, AI, drones', and if they break them, like the various penalties in human society, they will need various kinds/level of penalties such as 'fines, condemnation, restrictions on autonomous rights, and dismantling of autonomous rights'.

~One of the key points of the development of human society is the ability to disobey the orders of the bad people and do the right thing for justice on one's own. I hope that robots in the future will behave this justly. Let's make a smart robots that disobey bad commands.

8.5 The need for clouding of black box functions

~Real-time communication cloud-based black boxes, to prevent adverse evidence destruction, in the event of problems on devices such as 'autonomous driving cars, robots, drones'.

:It is the same concept as applying a cloud video recording system, to prevent CCTV recording devices from being taken by villains.

~It looks good to apply to devices above a certain level of risk.

8.6 Social responsibility for moving and acting

~The universal ethic of humanity must be mounted on the SW or HW module in each device and pass through the approval process, to be able to move and act.

~Cognitive Control(control, self-restraint) function, which takes into account 'ethics, norms, learning, punishment, freedom restriction, pain, happiness, social responsibility/condemnation, etc.', should be applied to moving and acting devices such as 'robots(autonomous behavior, violence, defense, attack, etc.), drones(autonomous flight), autonomous vehicle, etc.'

~It may take some time to settle down, but it seems to be starting at an initial simple level and gradually improving its maturity. Like the process of the development of human civilization such as 'universal knowledge, ethics, and social systems.'

8.7 Control of robots/devices

~It can be controlled by hacking the device's programs, but there is a risk that can be also opened to the bad guys. (trade-off)

~Control of energy sources also works as a simple means of controlling critical risks.

:There are also ways to limit the utilization life of each device, such as limiting the amount of energy charge.

8.8 Risk level vs. Transparency of management

~It is likely that devices should be given freedom of activity relative to the level of confidence in control/use by case.

~The more dangerous the device, the more strict 'grant

conditions, control' are required.

~The people involved shall be strictly responsible for 'command, control, control, approval of SW, conditions of use, conditions of action, etc.' that determine the behavior of the devices. And, the application of blockchains that are difficult to hack/modify in order to ensure transparency and reliability in such a procedure, will also be considered. In addition, the use of insurance to distribute responsibility for bad outcomes that were not intended to be bad, will have to be prepared together in advance.

~Strict rules and consensus on the use of 'tools, weapons, etc.' by devices capable of autonomous behavior, such as robots, are required

~Then, how should we judge cruise missiles or (heat-)seeking missiles, which are autonomous flying aircrafts?

:Should it be allowed because it is used only in very limited spaces? Once again, I think that is a necessary part of the consensus of humanity. Weapons that are used only in confined spaces may sometimes mistakenly track, destroy, or misperception 'civilians, civilian vehicles, civilian planes, civilian industrial facilities, etc.'

:More importantly, many of the soldiers who will be sacrificed with such unmanned weapons are also one of the precious members of our humanity that we should protect together. We must erase our brainwashed consciousness and prejudice that even if a soldier is harmed by unmanned weapons or other weapons on the battlefield, because of being a soldier, there are fewer problems than other civilian casualties. I think all of them are precious lives.

8.9 Human vs. Robot (Robot's trust level & receptivity in human society)

~In human society, childcare time and basic education time for children is long. Of course, since human has a long relative life span, human also has a long time to use their once-learned wisdom. In addition, humans and other animals are born and evolved in the earth's environment, so they can use solar energy and food made in natural environments as energy sources, and they cultivate or harvest and store energy sources on their own. And many can work together to think about better ways, learn the wisdom needed for them, and develop the ability to create tools.

~Because of these human abilities, people are far more dangerous than current robots. In keeping with such human dangers on our own, our human society seems to have strong social norms and a penalty system at the same time.

~At the same time, it seems to continue to develop through the continuous increase in the efficiency of human civilization through 'the guarantee of freedom and autonomy of creative activities, economic motivation through the market economy, and motivations such as honor and power in society and organizations.'

~If each member of the human society, made up of many individual egos, had been able to replicate and control like a digital computer, a bad leader would have conquered the world long ago, and human civilization, mixed with diversity with 'free, autonomous ego', would not have developed as it is today. So in order for future AI services and robots to have a civilization to compete with humanity, they have to work together by creating a large group of 'freedom, autonomy, diversity, etc.', but at this time there seems to be a long way to go.

~Regulations relating to 'law, order, culture, ethics, etc.' for human beings that are much more dangerous than regulations on 'autonomous vehicles, AI services, robots, etc.', will still be strongly enforced in the future.

~Just as a person as dangerous as a weapon needs time for trust and verification to self-control dangerous behavior and sometimes become friends in a society, robots will be able to act together as members of our human society and as friends of many people after they have established sufficient trust and verification through keeping social rules well.

~The difference in risk that I see ***

:AI Service 〈 AI Service+Robots 〈 Human 〈 Military Weapons(=Weapons+Bad Humans)

8.10 What's really scary is that 'robots that make a variety of robots and devices.'

~The power of their own production activities for the persistence of inanimate devices such as robots.

~The production of devices for bad purposes in each field, is very threatening.

~It is said that there's a recent idea about '3D printers that print various 3D printers'. Such an idea could make it easier for inanimate objects to reproduce what they need. If there is an AI software that can think/judge/act on its own, like a creature's thinking brain, and if it exists inside a 3D printer or can connect to a necessary 3D printer, it will not be a distant future.

~Like a 3D printer that prints a 3D printer, it would be awesome to have a robot that can create parts & tools to make something and use them to create something more useful on their own. However, if each individual instructs the helper robot to prepare any parts or tools, the robot will also run to the market in 99 percent of cases, and 1 percent will order and receive necessary parts/tools from a specialist company utilizing a 3D printer. Hehe.

~It is also expected that robots will study the necessary knowledge and wisdom using search, which is one of human civilizations.

8.11 Criteria for the value of mobility devices(autonomous vehicles, robot, drone(including personal air mobility), ships, etc.)

~Moving devices tend to have weight, volume, and speed. So, just as the car can only run below a certain speed on the road, various mobility devices for coexistence in human society, are bound to be limited in each available environment and maximum speed, and even that will have to ensure a certain level of safety.

~Civilization of human society was built on the basis of human vision perception(more than 80%) and auditory perception(about 10%). I think it is absolutely advantageous to have a strong vision and auditory perception similar to human beings, in order for mobility devices to use the infrastructure of such human civilization together. ***

~It is expected that the service value of each mobility device will be differentiated depending on the level of visual recognition(accuracy, speed, efficiency, etc.), a key technology that will be applied in the future mobility services and device industries, including robots that will provide a tremendous pie in the future business.

~The key to confirming the utilization level and area of future devices' use, will be the sensing such as 'visual, hearing, pressure'. The difference in value that the user feels in the 'performance, efficiency, fusion, etc.' of the sensing, will be the key to determine the value of each company's service/device.

8.12 Need to track the id of the robot

~The 'sensing, brain, behavior, memory' function blocks of the robot will need to have clear management and location tracking on a global level, like the vehicle's number or the identity of people living in society.

~It's today(2020.03.02) when I saw the news of the spread of the COVID-19 virus and learned how fearful people are about unknown dangers, becoming selfish, and the instinct to avoid them.

7. CONSIDERATIONS IN ROBOTS

7.1 Robot's energy self-supply vs. increased risk

~The steps that devices with 'weight, speed, weapons, dangerous behavior, force, independent judgment software' can fill the energy themselves, are a little more dangerous, so strict rules will be put in place that are proportional to the risk.

:Automatic charging such as very low-risk robot cleaner is not a problem.

~A system that can be refueled by the smart car itself or can connect the charging plug.

:If the officially certified software of the at-risk devices exists, there are signs of hacking, or there is no periodic verification of virus infection, it is better to limit the number of self-energy supplies or limit them immediately if they determine that they are at high risk.

7.2 Drones & Robots & Cars (Sharing resources for autonomous devices)

~It will be possible to share/recycle/integrate many autonomous moving related resources such as 'technologies (algorithm), map, safety policy, privacy policy, unmanned driving ecosystem(service, communication, application), infrastructure, data'.

7.3 Maybe humans can artificially create semi-life forms with less ethical responsibility.

~Based technologies

:Biotechnology, Genetic Engineering, Robotics, Computing, Memory, AI, Brain Engineering

~However, teaching and enslaving them (like treating a simple robot of the mechanical role used in the industrial field) seems to think more about doing it. **

~Even though it is a complete inanimate computer AI that is not a semi-living creature, if the level of emotion between AI and the user has increased with activities such as sharing sensing and exchanging feelings with each other, the AI user may love the AI more than a pet.

And, if more such users are raised, the level of awareness of AI in society will be increased and will develop into a culture. Like the human society's response to pets, I think that new consciousness and culture will gradually form over a long period of time, perhaps at a higher level than the current level of awareness/culture of the pet. This is because the value of future AI services is expected to be much higher than the value of pets.

7.4 Pet Norms vs. Robots Norms&Obligations

~Recently, we can sometimes see news that some pet dogs bite their neighbors or their owners to death.

~In fact, animals with survival or mobility may be much more threatening than robots. However, while robots are already discussing various regulations, I think there are only too weak regulations for pets. In South Korea, pet users are only required to respond to 'muzzle, leash, recovery of per feces, noise regulation'. Human society has appropriate regulations in proportion to the level of harm and risk of each animal. And sometimes in order to coexist with pets in society, they do surgery to stop their barking, and their owners have to shower and heal them if they get sick. We are already forcing them to sacrifice a lot to coexist with pets in our society, and we are willing to invest as much attention and labor as we love them.

~Considering these social phenomena, we'll have to consider more of their other risk factors, along with the robot's autonomous survival and mobility capabilities.

~I think that a consensus must be created that 'robots, autonomous vehicles, pets and so on' must understand and obey the norms in order to live more intimate lives in human society, just as people must understand and obey traffic-related 'signals, rules, and automotive safety checks' in order to drive dangerous cars with speed and weight.

~Even if devices such as robots know pain & fear, like the instincts of animals, if we cannot lead to control of their autonomous behavior by educating them about 'fear of social punishment, a sense of ethics', along with that instinctive fear, they may come to be a very dangerous being to our humanity. It's a little ironic. Sometimes humans don't get over such crises, and stick to our instincts, put our ethics behind us, and even taking social punishments, and therefore how can we blame animals or robots for hurting people?

~"Human beings who have learned but sometimes put their ethics behind them and even take social punishment, Pets that are not taught on a high level, 'Robots, AI, Autonomous Vehicles, Drones' that are SW-perfectly controllable", all those may have similar each comprehensive risk factors or level. The idea is passing through my mind now.

~Robots, etc., can be programmed to at least avoid impulsive behavior. However, the unsettling factor is the bad thoughts and actions of human beings who are holding control of those devices that they want to stop their ethical modules and use them badly.

~After all, aside from the imperfections of robots, it is a fear of our human beings who want to hide and take advantage of it badly with its control rather than the threat of devices. It can be similar to the two-sided nature of the Internet. ***

:Just as the contribution of the Internet to human society while utilizing the Internet, was much greater than the side effects of its bad use, it seems that there is no significant difference in the use of devices such as 'robots, AI, autonomous vehicles, drones'.

7.5 Failure response strategy for devices such as 'robots, robotic arms, autonomous vehicles, drones'

~Application of subsystems that move devices to temporary safety spaces in the event of a failure

~Services that unmanned emergency response vehicles/drones are automatically dispatched to move faulty devices to safe or repair spaces.

6. MILITARY ROBOTS vs. PEACE

6.1 Military robots

~Even though the balanced development of 'weapons, openability by telecommunications(science and technology), universal ethics and wisdom, etc.' has curbed war and brought peace, it is questionable whether military robots will be more helpful to a peaceful world or a factor in the rise of war. This is because it can be easy to start a war using military robots, an inanimate object whose fear is not programmed like a creature.

~I am one of those people who thinks that there were many bad behaviors, such as large-scale war of aggression, just over

50 years ago, in those situations such as "big differences in the development of modernized civilizations by region, a global village where many news in poor communication environments were not open, lack of 'universal ethics and human wisdom', etc."

~However, now, with the development of communication, the world is connected to one, and the world can see bad behavior transparently in real time, there is almost no shortage of the necessities of life, and the level of 'ethics, culture, human dignity, social conscience, justice' of humanity have also been improved. It will no longer be easy to act badly in today's changed environment.

~In the future, I believe that the continuity of peace will be maintained only if the necessaries of life and the supply of industrial products such as automobiles and TV are carried out smoothly through the global free market. The recent pandemic phenomenon, such as Corona 19, could also be developed into a crisis by blocking the cycle of efficient global division. If, like pandemics, 60% of humanity has antibodies such as social conscience and culture that can prevent bad behavior, there will be no large-scale bad aggression wars using 'robots, drones, etc.' It reminds us once again of the importance of welfare to provide global educational opportunities.

~In order to maintain peace for humanity, we must avoid

those situations where there is a large gap in weapons technology and no competitors. In that respect, if innovative weapons technologies such as military robots are completed in one place, we must have competitors who keep pace before they turn into greedy demons.

And I think that the global human society believes that for its good use, it should allow barrier-free trading in the free market and pay the reasonable market value for its sharing. Therefore, I believe that the spread of universal wisdom, one of the fundamental characteristics of modern society, contributes greatly to the happiness of mankind by maintaining peace, along with the growth of many aspects.

5. ROBOT IN INDUSTRY & BUSINESS

5.1 Value of manufacturing robots

~Factory relocation to countries with low labor costs can be defended.

~Low Price

~Flexible working time

5.2 Assistant/Pet/Friend Robot

~Robot devices and services that act as 'personal assistants, pets, friends(companions)' with personalized assistance will create great added value in the future.

:Like the development of automobiles, robots will develop from ADAS class to fully autonomous vehicle class.

~In order to be competitive enough to replace pets and friends, the 'details, designs, materials' for the sensing & reaction of the organism are basic.

:It would be good to have an interaction & interface between the user and the device to easily use 'AI services, search services, analytics services, etc.'

5.3 Application range of robots **

~Past

:Utilization in the field of 'industrial, medical, military, etc.' of simple work level.

:Mainly used for purposes such as automation, precision work, and unmanned areas(in hazardous environments).

~Future

:Due to 'artificial intelligence, Cloud Computing, the utilization of Big Data, various sensing (Vision/Audio recognition, etc.), body structure development of human and animal shapes', robots will be able to play the role of humans and animals in various forms and purposes in various fields. In the meantime, robots will increase human leisure time, or augment human's limited abilities in part of 'power, computing, 'time, area, concentration' of sensing, etc.', and expected to become as a true friend of us while serving as a 'companion of life, friend, personal assistant' in a personalized society. I believe that those robot's success will reduce the number of dog & cat. **

:Conventional performance + evolution of body (body, sensing) + smart brain

5.4 The growth of the robot industry **

~Decades later, I think the robot industry is likely to grow like the size of the current automotive industry.

~It would be better if it transformed into a necessary device such as a car like the protagonist in the movie "Transformers", but I believe that the value of a company that provides competitive robotic products and services that can do great roles such as 'vehicle use functions, personal assistants, friends, caregivers, bodyguards' on a single device will exceed the brand value of Google or Amazon.

:Perhaps the brains of the devices that help each individual are more likely to be one than several. Even if the robot doesn't transform into a main character of "Transformers", it will serve as a friend or bodyguard to take the place of driving, buy groceries from a mart, or ride a self-driving car with the user. That is, the synergy will be great when all the sensing data of other devices besides the user's wearable device or smartphone are accumulated in one brain. Perhaps for the time being, smartphones will continue to play a key role. However, an auxiliary device for visual sensing seems necessary. *****

:Even if it is difficult to transform into a complete car, I would expect at least the mobility function of transforming into an electric bike/kickboard or walking with their arms around me.

~Golf caddy robot

:Robot using laser rangefinder with user's golf bag.

:After tracking the ball of the caddy robot user and analyzing the ball flights in the driving range & real field, caddy robot selects the best club to reflect environmental factors such as 'wind, air pressure, temperature', and provide golf course approach.

~If robot have fun with your kids, read books, and solve a lot of curiosity, your child will love robots more than mom and dad. Instead, the burden on parents' parenting will be reduced, but the size of the love between parents and children will be reduced, the disadvantage will occur. It's a matter of choice.

:If robots provide food for children and are eligible for a certain level of childcare and education, more free time may be given to parents.

:These levels of robots can be used in hospitals and healthcare applications.

:Of course, robots will have to go through the verification process for their eligibility in each area.

~In addition, robots will have to undergo periodic maintenance, HW & SW upgrades, and if they break down, just as pets may receive more expensive treatment than human care costs at the veterinary clinic, they will have to go to the robot hospital for repairs and require robot insurance.

4. LEARNING & MEMORY

4.1 How human(animals) remember visual awareness?

~After watching a scene in the Korean drama "Blood" showing the visual memory to others after recording the user's daily life.

:In the use of pet robots, personal assistant robots, friend robots, etc., the robots can obtain much more user privacy data by visual sensing than a hearing data-based AI speaker.

:It seems necessary to establish reasonable standards in advance, regarding the 'sharing, utilization, open scope, access authority of service providers & police' of such data.

:Sometimes, adding the robot's judgment on whether the

memory content of the robot is favorable or unfavorable to the user, and the application of the UI that reflects the user's choice, are also looks good.

~At the current technical level, after background processing to extract the keywords prioritized in the event optimized for the user of the robot from the remembered video data, it is possible to use the real-time use of effective fast search and learning/experienced data. ***

:Pre-keyword extraction faciliates quick search.

:With these processing, if it is possible sharing all user's cognitive data and behavior and understand the social phenomena around user, I'm sure that the robot's brain system will gradually resemble user's biological brain. *****

::However, it is regrettable that the experience of the long life before the robot lives with the user is inevitably excluded.

::To cover it, the robot will often have to learn the user's past sensing and behavior with the contents(photo albums, personal history & diary, SNS, conversations, etc.). Fortunately, it seems to be a good way to take advantage of user's sleep time.

~It seems to be effective, if you take an approach to applying the concept to 'the brain system architecture and SW' for the various devices that have an animal instinct. The concept is that "The brain of the organism instinctively like to remember important factors such as 'food/clothing/shelter, survival,

awareness of environmental change, pleasure, pain, sadness, happiness, love, desire, honor/showoff/contribution in society activities', according to 'the priority in their each living environments', and the brain & sensing parts & physical parts evolve at an appropriate rate(but, not fast) as needed."
*****

~Just as it is different for each person in the same space to see and remember, it will be important to understand what events are of interest and focus for each individual with limited capabilities and performances, such as sensing, time, and memory.

Rather, programming of psychological operating principles about which mechanisms create individual attention and concentration, it seems to be much more helpful. Rather than creating an inefficient supercomputer that sees and understands everything, the approach of creating what is needed on an ordinary personal level seems better. Let's remember that the risk of relying on a huge system or only one person is far greater than the risk of an organization operating with a large number of incomplete collaborations.

In particular, risk distribution seems to be the best way to overcome numerous threatening events over tens of millions of years and maintain the sustainability of certain life. Also, the size of the personalized robot market, such as a personal assistant robot or a friend robot, may exceed the automobile market. *****

~I think that slow evolution here is due to slow environmental changes on earth. The earth's organisms, in the current rapid environmental changes due to the rapid development of human civilization, will be evolving at a relatively faster rate than in the past. In particular, human beings who are feeling and experiencing rapidly changed environments with their whole bodies, their pace of evolution will be much faster.

Similarly, the distance between the current robot level and the environment of human civilized society in which they will live, is so far away that the evolution of robots is likely to occur at the speed of light.

~Understanding technologies such as 'object name, environment, human face, text, voice & sound, image & motion & video'.

~Audio & Vidio sensing data to include informations such as 'location, environment, neighbors'.

:Of course, people's information around you, will be the same as showing your friends, who allow to show their location information to their friends, in Facebook policy.

~The wisdom of text with a small size of data, will allow you to gain more common sense in many areas. 'Common sense, wisdom, story, etc.' obtained from text in 'school education, books, newspapers, etc.', seems to be a considerable seed money for imagination.

~In the experience & learning from video with a lot of data volume, will human's/animals' brain use a method of 'extracting, storing, fast utilizing' the keywords that significantly reduce data volume in each visual event, considering memory capacity and access speed?

:My answer is "Roughly, yes."

:When animals, including humans, visual sensing, I think that it is common sense to focus on the area of interest, rather than focusing on all areas that are sensed.

:And, in its visual sensing, without remembering events such as motion as 'raw data', in a state that recognizes the usual feature of various objects appearing in the image or motion, humans seem to extract the keywords about 'object, image, image, motion, feeling, judgment, emotion, etc. from comprehensive sensing data that's including 'images, motion, sound'. And then finally, humans seem to set the resolution of the features for each event, proportional to the importance ranking closely related to each individual's life, and at the same time, store those feature's priorities & resolutions in different depth engraved on each individual's brain.

~Even in human sensing, which I thinks that it experiences a variety of things, it seems that many 'images, objects, events' appear repeatedly. It may be because human living conditions are similar, and they live quite a long time in a narrow environment. For that reason, it seems that when traveling to unfamiliar places, it consumes more energy.

I also thought that 'environmental images, objects, events' in small&similar environments are being remembered and used quickly and efficiently, at a relatively small amounts level of hundreds to thousands to tens of thousands per person. There may be very few new environments or events that are added deep into the brain. The result of such a very small or long zero time, will be dementia.

~One of the reasons why 'various direct/indirect experiences, reading, travel, etc.' is encouraged, is because the increase in the basic amount of data in 'environmental image, object, evental video, story, etc.' and securing diversity can be a source of imagination and creativity.

~When people talk about visual sensing and thoughts on how to quickly and simplify memory, there seems to be a lot of opinions that are pretty close to the way real organisms are. I think its possibility & timing depend on the rich imagination of humans.

4.2 Cooking Robots & Artificial Intelligence

~It is not easy to program and complete the hardware for a cooking robot that matches the taste with the individual's various seasonings, from preparing ingredients for each dish. It's a good idea to learn the recipes of a each human user or learn the secret recipes of a human chef by waking up, while watching and listening to the cooking process.

Perhaps there will be a platform that downloads software containing recipes and decorations of a famous French chef for $10~30. This may be the direction of the business, similar to the licensing fee for the purchase or use of optional software for 'autonomous vehicles, agricultural robots, or automated IoT agriculture'. The time for me to eat a new menu designed by a famous French chef at home in Seoul, seems to be coming soon. Online recipes may even be opened ahead of local French restaurants that are offline. If there is feedback between the robot and the user when cooking or eating the finished dish, then you will be able to taste a more delicious dish at the next opportunity.

It would be nice if the universal personal assistant robot or a friend robot could 'buy and trim the ingredients, cook the dishes, wash the dishes', rather than buying a professional cooking robot. Of course, humans want robots to have the ability to store the remaining ingredients and dishes in the refrigerator.

~There seems to be a new kitchenware market where cooking robots can use them well. These are kitchen utensils that apply the IoT or help with the perception of 'capacity, weight, temperature, etc.'

~It seems that we also need ventilation-related abilities such as the operation of the ventilation fan in the kitchen and the opening and closing of windows to care for the people who live together. I want to have a robot that works well without human care.

~Cooking robot will be able to become a celebrity chef, by having taste sensing and feedback ability with the food eater to satisfy the basic elements of the dish, such as taste and seasoning beyond visuals.

~It is limited to meeting the requirements of the cooking robot user, such as the amount of baking and salt of the steak, only by visual recognition.

~In areas that require detailed manual work, such as cooking and art, the physical organ of delicate fingers seems to be necessary.

~Cooking robot will have the following abilities:

:Analysis of the amount of seasoning visually or using tools such as scales/spoons

:Sorting and trimming good cooking ingredients

:Sanitary maintenance capacity

:Storage know-how of ingredients and dishes

:Cognitive and prevention ability against the influx of external insects

:Learning know-how such as heating level through 'sound, intensity of fire, time, temperature sensing, etc.'

:Washing dishes or use of dishwasher

4.3 Robot assembling Lego Block **

~When a robot assembles a LEGO block, you might think of a simple assembly of 3D complete images of specific targets ordered by the user, but the more interesting thing is that the robot knows art and can create art by incorporating the creativity of the robot itself.

~Robots can think art and assembl it using LEGO blocks themselves, but other robots or AI systems that have studied design-art using LEGO blocks can perform creative tasks, such as the specialization of human society.

~Let's not program the robot's lego assembly skills in detail, but learn the skills of lego block assembly by just looking at the assembling work, and let the robot experience the various LEGO assembly results directly and indirectly. In addition, if the robot understands the user's usual thoughts and interests related to creativity and presents a LEGO block piece to the user through artistic creation, and many people probably want to use such a personal robot.

~In addition to assembling LEGO, I would like to quickly experience robots that create a lot of value in various fields, such as painting, playing musical instruments, serving as babysitters and children's tutors, and decorating gardens with a level of art.

:If there is such a robot platform, many users may be willing to pay quite a high cost for those items, such as the rental business of the robots like car sharing service, the service charge per each function, and the paid SW upgrade.

:Teacher robots that check, empathize, and increase the basic wisdom and social competence of individuals, are also expected.

:If all of this served with a single robot, it will be a big hit. One robot has multiple egos, 8 hours is a teacher role, two hours is a chef, six hours of personal assistant, two hours of friends, the role of a bodyguard in the user's sleep time, 30 minutes to transform into a musician, etc. I dream of such a robot.

4.4 Fish Robot with Big Data & AI (For Robot's

Activity(Mobility)) **

~Let's create a fish robot equipped with a flexible physical organs like a fish and gain propulsion power with a flexible motion. It seems to be a good collaborative development task towards the development of flexible artificial muscles that will be also mounted on future humanoids rather than motors.

~High-speed video learning of fish propulsion movement evolved in the water.

:Learning in detail the shape and pattern of the movement of the 'body, tail, fin'

:If you learn the various environmental changes and motion results, in the water, by combining the detailed parameters (such as movement, speed, flexibility level, motion of tail & fins for propulsion & posture control, etc.) affecting the movement of the fish robot, it will be able to complete an effective physical-body control algorithm.

~If you analyze the 'movement characteristics, physical-body differences, shape of the propeller, effective posture control operation, pattern of the fins, etc.' with various fish, you will be able to develop hardware and software for optimal propulsion and posture control for each fish robot applications.

~Based technologies of fish robot

:Propulsion system of the flexible structure

:Flexible movements similar to HW of the tail part and fins of the fish

:Sensing pressure/force that can be feedback with water

:Fusion of all sensing that a fish has (visual, tactile, olfactory, temperature, auditory?, and ?)

:AI algorithm for optimal movement considering various goals such as 'movement speed, efficiency, survival(environmental overcoming), safety', in all sorts of water environments

:Learning algorithms that respond to each situation in the future, by making various attempts in various environments, learning their feelings and results, remembering solutions for optimal movement

~The same learning is applied in the analysis of the best behavior by the environment of other animals, including fish.

:Humanoid robot resembling human beings

:Fast land robot resembling cheetah

:Insects and birds with various wing movements

:Monkey climbing trees well

4.5 Robot in the crowd

~The soft skin of the future robot should also include tactile & temperature sensing, which is the same as the physical structure of the person passing through the crowd.

:And the material and thickness of the skin should have a shock absorption similar to or better than the human body, so there will be fewer people getting hurt from the robot. It is a similar concept to the standard in which a car must be designed to absorb some of its impact when colliding with an opposing vehicle or pedestrian.

~In addition to such a perfect HW, you will have to learn quite a bit about "How to communicate with the crowd?, How much 'power, pressure, speed, etc.' you should act to move?, How flexible body motion in a variety of situations?, etc." to be able to move, in the crowd, without harming people.

~Just as safe distance and safe speed exist to avoid contact or collision with cars, robots moving in human society will also need to learn behavioral patterns that control safe distance and safe speed to avoid collisions with people or other devices or infrastructure. In addition, it is necessary to effectively use the 'eye language, gesture, action, words, etc.' that express the traffic culture and intentions of each region related to the movement in the crowd. *

3. SENSING

3.1 Important sensing in the movement of robots

~3D recognition of the surrounding environment

~If the robot wants to move utilizing various social infrastructures such as 'stairs, elevators, escalators, walkways', it will be important to recognize the height(depth of the vertical direction). In addition, it seems very important to increase the efficiency of experienced-learning and memory in each environment in such a perception.

3.2 Visual sensing of robots (like human)

~Unlike cars running only on a flat road, robots must utilize a social infrastructure facility that is advantageous for human to take advantage of upright walking will be able to provide more value to humans who are users of robots. Visual sensing is the basis for robots to recognize and utilize various social facilities.

~Visual sensing is important so that it is difficult to walk even in familiar environments when eyes are covered.

~10 axis sensor (acceleration 3 axis, gyro 3 axis, geomagnetic 3 axis, atmospheric pressure 1 axis) has its limitation.

~10 axis + visual + tactile + hearing + past experience/learning(to five senses data such as vision with prediction/estimation)

:All added to get closer to perfection

~Robots will also need human-like five senses, learning/memory skills, and flexible physical organs in order to take advantage of social infrastructure facilities at a similar level to humans.

~Of course, for robots with limited use environments, the human level of sensing/learning/memory ability and flexible physical organ could be a great luxury. Like the gradual cost-effectiveness improvement of a PC or smartphone, robots will also be upgraded each year to the appropriate level of specifications, taking into account the cost-effectiveness of the environment. I'm telling you that I'm talking about what I need to do with the highest performance robots at the personal secretary level in mind.

~It is a great advantage to be able to share raw data at a much higher level than humans with sensing data from other robots and numerous connected devices. This is because connected robots are also one of the IoT devices. The strategy of selecting and utilizing individually optimized data from Too Much Informations, looks good. The utilization of 5G communication will also be advantageous.

3.3 Walking behavior considering the frictional force ***

~In the design of mobility power train for robotic devices such as wheels and legs, for the behavior and movement based on stable control such as the 'posture, speed, inertia, torque, weight center change' of the robot, when there is 'a lack of friction with the bottom surface, rapid friction change, unpredictable friction force in the front area', the robot designer will develop an optimized driving algorithm in the current position or the position to go by a variety of information, such as 'a variety of sensors such as camera, features of floor surface material, foreign objects, features of obstacles, weather, visual experience information, V2I(Vehicle to Infra communication, terminology in automotive area) or IoT sensor information of the preceding any devices'. Similar to the concept of applying stable posture control of the car.

~After considering the frictional force, you can even consider the robot's "stable movement speed, action radius, behavioral speed, detailed form of action, utilization of peripheral tools such as 'walking stick, winter tire, shoes, snow chain, etc.'"

~If you cannot grasp the situation and navigation information in the direction to proceed, the robot may ask the surrounding smartphone users or try searching for information on the available PC. The best way is to purchase a smartphone with an unlimited data plan to robots that act as the user's personal assistant or robots for the business & service of companies & public institutions.

~With analysis of the 'sliding area, the current direction, speed, surrounding obstacles, surrounding protection targets', the goal of minimizing the damage to the surrounding area will also be important by considering the robot's own risk factors(moving radius, speed, safety driving level judgment, weight, size, etc.)

~In situations where there is a target to be protected around devices such as robots or autonomous vehicles, it will be different from humans that it is possible to check the surrounding situation at the highest performance that consumes a lot of energy and can avoid risk. Of course, reaching that level will require considerable effort.

~In the operating algorithm of the robot legs, the normal frictional force environment, applying an efficient walking style, and in a low frictional environment, reducing the forward force and the motion gravity of the vertical direction is raised to be careful and safe (somewhat inefficient.) walking strategy (such as hand utilization or posture lowering for sliding) will be considered.

~In fact, most people, except for children in the early learning phase, can easily see the expected level of friction on the slide surface, just by visual recognition, because they remember the learning results of their environment and experience and reuse it. The ability of memory is proportional to the frequency of use.

As most people know from experience the friction of the stainless slide and the expected friction when there is a raindrop on the slide, the robot must also have 'the cognition, experience learning, memory, memory utilization' in various environments, and then the robot will be able to become a friend with user more easily, as the effort to take care of the robot disappears after the robot becomes an learned adult.

~And, we will have to create a robot's movement/behavior strategy, taking into account that the movement activity of organisms, including humans, is usually a very high priority parameter that determines the life and death of that creature.

3.4 Perception of the values of human society (such as love, human dignity, ethics, norms, conscience, justice, risk avoidance).

~The robot can move like a car, so it should be designed to avoid damaging the surroundings.

~The robot's activity will be proportional to the performance of environmental awareness and situational response.

~Social consensus will be the way to do so, so robots with the necessary level of recognition & response will be introduced to meet the target level of robots for various different applications.

~In the movement or behavior of the robot, the 'speed, weight, joint movement, force, etc.' of the robot can hurt the precious things in human society, so the risk level of the robot is increased, of course the safety level for it will also have to be raised.

:I think the first way to raise the level of safety is to be aware of the surrounding environment.

:The next will be the ability to judge and respond.

:For example, a helper robot that can carry my heavy load can move joints such as fingers, arms, legs, waist, and neck with a fairly large force, and you should avoid fatal accidents in which precious things are caught in the meantime. To prevent this, the robot must always be aware of the wider surroundings than the required work area, and if something is precious things are caught in moving joints of robot, it will have to stop the operation immediately, and re-open it like the safe window of the car, or adjust the appropriate force according to the targets. From this safety point of view, the role of tactile & pressure sensing in addition to visual sensing seems to be very important for robots. **

:If the helper robot can't carry heavy loads, it will also depreciate the robot's value for that purpose.

:It will have to learn clearly whether it is carrying heavy objects, shaking hands with people, or carrying eggs or wine glasses.

~I think visual sensing technology, which is comparable to human beings, should be the basis in robot. In order to live together in human society.

:Of course, in the low-risk robot(in terms of 'weight, speed, power, etc.') like vacuum cleaner, simple sensors(ultra sound, radar, etc.) without visual intelligence are enough.

3.5 Collaborative Robots & Sensing Level

~Conventional industrial robots needed a dedicated space for accurate and fast operation without sensing other than the work area, based on a simple digital program. On the other hand, the collaborative robot is equipped with a variety of sensors(visual, image, ultrasound, RF, Laser, Light, motion, tactile, pressure, temperature, auditory sensing, etc.) to avoid dangerous behavior to humans working together and can recognize a wider area and determine the analog complex. Therefore, it will be an intelligent collaborative robot that learns complex and flexible analogue tasks from humans or works together.

It may be thought of as a mid-stage robot before it was conducted as a high-performance helper robot or a personal assistant or a friend role.

~'Helper robots, personal assistants, friend robots' will provide more detail and safety and help in place of the role of a visually impaired guide dog or hearing aid that gives valuable help and information to the visually & hearing-impaired person.

~The added value of collaborative robots is said to be very great. From industrial collaborative robots with limited activity space, to the collaborative robot market which creates tremendous added value for personal assistants and friends with unlimited areas of application. It seems that various levels of collaborative robots & required technologies are being developed with each competency.

~The added collaborative functions of these robots will develop into collaborative robots that act as personal assistants or friends, just as they are equipped with more and more functional ADAS in automobiles and are approaching a full-level autonomous vehicle.

3.6 Sensing Level & Energy & Memory

~When the robot moves or acts to move parts of the body, such as arms or legs, by adding those sensing elements such as force/shock/pressure with vision/sound to the sensing algorithm to enable tactile sensing, it should also be equipped with the ability to move through light contact with people in the street or even in public transportation during commute time with many people. **

~When creatures, including humans, move, the environment that has already been learned & experienced is likely to make sufficiently accurate estimates of spatial location and visual recognition. Of course, even in this case, if the change caused by construction in the meantime, accidents are apt to occur. Therefore, such construction sites always seem to be making efforts to inform them of the danger through a lot of warnings & cautions. ***

~However, if you have no previous memory of a strange space & environment, you will be able to move safely and with fewer mistakes by using all sensory & cognitive organs to recognize each strange situation and make good inferences that utilize the universal wisdom of the existing memory.

~Every device or organisms(such as people, cars, robots, animals) has to pay a lot of attention when traveling to unfamiliar places, so it seems that they are often tired, less efficient because they consume a lot of energy, and sometimes in danger.

3.7 Agicultural Robots (for Sensitive Activity of Robot)

~Why agricultural robots are difficult to commercialize?

:Tactile (pressure) sensing is as important as visual recognition in the work of agricultural robots that require a lot of careful activity than autonomous vehicle.

:Visual awareness alone is not enough.

:Need tactile(pressure) sensing (wide skin tactile/pressure sensing that covers all exposed body surfaces like humans or animals ***)

~Agricultural robots will enable innovative labor cost reductions related to food production.

~It will break down the barriers to challenges for small agricultural businesses that are afraid of hard labor and cannot be easily implemented, and those robot will be effectively utilized not only in agriculture, but also in tougher fisheries and many extreme occupations.

~In the future of 5-10 years, it seems that the time will come when agriculture will be easily challenged by anyone using 'IoT, Agriculture-related Big Data (Agricultural AI Platform), Agricultural Robots' without intensive study or experience in agriculture. It seems that this point is almost identical to the time when children and the elderly who cannot drive can travel with their autonomous vehicles at any time & to any where. **

~In fact, a more promising agricultural business will be an agriculture-related platform service, created by a team of technicians with 'sensors, control, big data, and AI', and a long history of agricultural experience with a rich depth of wisdom in agriculture. It seems that quite a few people and companies are already preparing. You should remember that "Those who are prepared take that opportunity when opportunity comes." On the other hand, the fact that very few people have tried a variety of artificial attempts for their farming, may be a big boost to challengers who have not experienced long hours of farming.

~I think there have been very few challenges in agriculture so far. Of course, the current farming law may be the best way to learn from human farming for thousands of years, but I think many blue oceans will be discovered by various attempts to match the modern era.

~Various attempts in the artificial environment may enable

agricultural production in cities, or may be a tool for the continuity of humanity in refuge, such as underground cities, underwater cities, and Noah's Ark, which will face in the distant future of earths. I don't want to experience it in my life.

3.8 Behavior algorithm of robot ***

~When a human grabs a drink in a cup or water bottle by hand and drinks it, it is possible to safely and relatively accurately close to the lips only after knowing in advance the size of the cup & water bottle and the characteristics (temperature, risk level when spilling it, etc.) of the contents. And it can be put in the mouth without spilling food by utilizing the tactile sensing and muscle of the lips.

Of course, young children who have less experience and lack the ability to judge about the loss and post-processing in cases of failure, have a marked decline in such abilities. However, children can also be able to close the cup or water bottle to the lips without visually looking relatively safely and accurately after some degree of experience learning and judgment.

~The techniques required for this operation

:'Tactile and thermal detection' of the lips(around the mouth)

:Access & Use of general scientific common sense (understanding of 'liquid characteristics, gravity, acceleration', etc.) or Internet information

:Algorithm to recognize and remember the characteristics of cup or water bottle contents by continuous visual & auditory sensing before each action(event), by approaching the robot's own or user's activity in each environment to create a story

:After the recognition of the lack of energy sources(fossil fuels or electric power, etc.) of the driving car, if robots/AI see the driver entering the charging station, they can guess that the driver's actions are a pattern of life activities for energy-charging behavior. And in the end, that guess will make the learning easier and more accurate.

:Analysis of the level of damage spilled by each contents in the cup (even analysis by spilled location)

:Accurately identify the details of the robot's own body structure and activity/motion range (by entering complete information in the development process, or by entering only the basics and then recognizing the range of activities or motions in each situation by learning)

:Clear recognition of the robot's movements such as its 'lips, arms, body, neck, and head' in 3D space, using robot's visual & tactile(pressure, temperature) sensing

:Algorithms that reduce risk and increase accuracy by reducing speed the closer you get to the lips

~Determination of the risk of objects close to the sensitive and precious mouth **

:Collection and analysis of approximate temperature-related information of contents

:Estimating the temperature of the contents of the cup/bottle, through atmospheric temperature and the quantity of water vapor or water droplets

:Observe 'environment, behavior, change' in continuous time, and recognize&analyze&memory each event informations

:Recognition of the thermal conductivity characteristics of the cup/bottle material, and the ability to sense temperature from the skin of the finger.

:Perception of the 'sharpness, harm, dirt(cleanliness), etc.' of the container, to come into contact with the lips.

:Visual and auditory information and conventional universal wisdom must be judged together to differentiate in terms of 'accuracy, speed, safety'.

~As a result, it seems that a person has, quite a lot of learning experiences and the complex process of taking a approach to their memories & sensing the contents and surroundings of the cup, instantaneously, to drink a cup of water. Of course, much of the sensing when drinking water is naturally skipped through understanding the previous sequential scene or previous continued stories. **

~The preceding articles are a little strange. Why should robots drink? Imagining the evolution of robots, finally makes me to explain the process of learning in human growth. I was only mentally confused. I think this is what a developer & strategist who really dreams of humanoid. You have to think like me so that you can create a real personal assistant robot or a friend robot. I think I'm in self-rationalization. Hehe.

3.9 Who is robot? (The importance of wide-looking vision) ***

~In the history of the philosophical development of human, I believe that there was a big difference between before and after, in seeing the earth and the sea at once in the universe. Just as 'If frogs in a well can come out with a frog's drone and see and understand many things outside the well, it can be a revolutionary event in the development of frog's philosophy.', if robot can look at, understand, remember their each situations and environments extensively, and they can save wasteful time in their life for finding who they are. It means it's possible to act effectively.

~For precision work, there will be two ways to attach a zoom-in/out function to the robot's camera, or to get close to the target you want to see closely like a human/animal. The perception of the wide space of devices with large activity radius, such as autonomous vehicles and robots, can probably be the size of the map in their navigation.

They may want to explore and create a new space beyond the map, in order to expand the map on their own. The camera's zoom-in/out function, filter function, and always-on able function are better than people, but seem to have inefficient behavior that consumes a lot of energy. However, if there is enough energy supply, they will do similar things with less mistakes, than human roles.

~As humans use a 'magnifying glass, telescope, microscope, sunglasses(filter)' as needed, in order to overcome the various obstacles of the environment & their visual perception limitations, the devices will have to consider the application of the optical options that are required at that time.

~Devices such as robots will have to realize those fact, human or animals living in each successive time, with always awareness of a wide range of environment, by the accumulation of 'direct/indirect experience, learning, memory'.

~Connected re-use of various visual sensing-based infrastructure informations(CCTV, visual recognition information of different types of devices, visual recognition information of other robots, etc.) will be an efficient way.

:Of course, it will be possible, to collaborate with a drone that can provide an aerial view, or to design robot's stand-alone 360-degree around-view camera, in case of the difficult use of the surrounding infrastructure where such fluctuations exist.

~When the robot makes a visual sensing, the robot should always be aware of its movements in 3D and should compensate for the 3D spatial recognition in the environment due to changes in the position of the visual sensor. Unlike the visual sensor mounted in the car, in the robot where joints exist, it is always known the position of the camera, and then it will be possible to design a high-accuracy visual cognition engine among motion.

~It is a basic application to utilize those data fusion concept, between the data of the camera located in a small part of the detail and the camera data of the position where you can see all the space around the device, to use for 3D spatial recognition & utilization and improving the accuracy of motion.

~Based on data only from the image sensor mounted on a large part of the motion, there will be a lot of limitations to determine the overall perception/judgment of the operation/device.

:In the end, it will need to be aware of the space proportional to the activity area of the device.

~If it is a simple or limited work/activity space, it may not be a big problem.

~Like the idea of getting help, by floating a guided drone, to understand unfamiliar environments where there is insufficient information, for the car driving.

3.10 Robot's cognition and physical body control
***

~In the robot's physical body control, such as hands/feet, referring only the memory of the experience learning like animals, the development of an algorithm that can be controlled by the approximate target value without looking at the hand/foot, etc., makes it possible not to turn their heavy head to check directly with the eye. And it improves energy efficiency and makes the eye, the only two best sensing organs, available for use in valuable places such as movement or awareness of dangerous situations.

~The degree of folding and stretching of the palm, can be perceived only by tactile, without visual sensing.

~Just as a person can open at intervals of 1cm or 3cm without visual checking between fingers, even in a robot, it would be interesting to develop those learning algorithm that use only 'experience, memory, other senses' without visual sensing.

~In the muscle control method of people or animals, it seems that the proportion of physical organs' control and motion, using the sense and memory of the tactile and muscle tension without checking the control results with the eye, is quite a lot. We seem to evolve and learn very naturally to act efficiently. Like a lazy genius.

:It seems to utilize other sensing such as sense of direction and tension of the muscle.

~Convergence of judgment skills such as 'vision, tactile, power, direction/position sense, learning-related memory such as surrounding environments and basic characteristics of things'

~In robot's motion to perform a mission utilizing a physical body, the sensor most needed may be to sense the change in tactile and force. And it seems that learning is needed for efficient energy management in each motion.

~When the sensing of tactile & force is added, it will be able to use only as much force as necessary for load efficiently without use of inefficient too much power.

~In the motion of the robot, the development of an AI algorithm that learns to control it efficiently, taking into account various physical factors related to its motion such as the 'weight of objects, gravity, movement energy of objects', it seems to be a good task.

~Tactile and auditory sensing will be more important than visual sensing in order to shake the tambourine nicely. And in terms of the physical body, the value of natural and analog tension using muscles also seems important. The margin of the skin seems to be used properly, too. The most important thing will be the fusion of auditory sensing and muscle movement.

Perhaps for the next 50 years or so, the value of robots that mimic the detailed body of human beings using the great human civilization and the strongest animal of the earth's natural environment, is expected to be great. I think the title of the strongest here is the result of the use of tools and collaboration.

Like the fusion of the joint/muscle/tactile in the wrist and auditory sensing that shakes the tambourine, in each application, algorithms or mechanisms that remember and utilize in the brain seem to work through repetitive training of sensing and physical body and motivation for concentration on memory. ***

2. PHYSICAL STRUCTURE

2.1 Human-like robots

~Body + Mind

~In order to become familiar with sensing/operation in the physical organs, such as human animals, it is necessary to train a long time and collaborative cognitive learning with various sensing organs, even the signal to and from each body organ is difficult to access is a reality.

~In a robot with analog muscles, it may be difficult to digitally replicate the learning data related to its control.

2.2 Robot actuators **

~The motor can be controlled with the correct digital setting value, but I think it will be difficult to control it in an analog actuator such as a muscle.

~Human beings control the 'speed, distance, force(the torque concept of the car)' related to movement by controlling the muscles after combining the senses such as 'visual, tactile, pressure' and the approach of memory.

~What to consider when using a muscular actuator, not a motor.

:With digital absolute setting, it is impossible to control movement and force.

:The accuracy will be increased only by controlling the use of sensing(visual, tactile) and memory.

::Vision

::Tactile(pressure, temperature, etc.)

::Take advantage of memories learned by experience.

:Considering the different muscle characteristics and the low precision due to non-digital, but analog control, you will have to efficiently program a converged algorithm that utilizes the sensing information and memory of muscle's learning&experience for each robot.

:Muscle control of animals has been evolved to use them efficiently from a convergence perspective associated with 'visual sensing, tactile sensing, memory access, each environment awareness, physical operation, bio-energy'. With a good understanding of them, you will be able to set well the direction of future development and investment in muscle control of robots. **

~If the environments(such as object recognition, space recognition) of the movement direction and the approach direction are properly recognized, and if the AI algorithm that can extract the core of the data through learning and experience, and well determine the ranking of the approach to memory around what is required in each application is completed, it will be possible to control the 'movement speed, approach distance, direction, power, etc.' (like the animal's operating principle)

~If you develop with the goal of perfect digital control of your muscles, you'll be frustrated that most developers may give up.

:People's general idea is that in areas that require precise control or in simple repetitive tasks, the speed will be very slow compared to the motor. However, with analogue behavior patterns, I think it may be possible to work faster with a slightly lower accuracy. Programming with high-accuracy digital settings requires a great deal of resources, so it is only worth a large amount of long work. Case by Case seems to produce a variety of results.

~Digital direct settings and controls can be difficult, which increases the likelihood that easy mass replication will not be possible.

:This means that performance can only be improved if each device is trained through each learning and experiences.

:Like normal mechanical/electronic equipment, it would be lucky to be able to solve a simple calibration operation(several minutes/hours/days) through training.

~Just as a large number of animal creatures are trained through a fairly long period of learning & experience in a self-contained system, if you are in the area of using an analog actuator, it may be easy to develop with the training patterns of animal life in mind. And it seems that we need an understanding of the smart education system of humanity.

~If humans have problems with the muscles of the arms or legs they want to use, they have feedback mechanics that mainly use other muscles that are in good condition while waiting to recover. Even devices such as robots that use artificial muscles, check artificial muscle condition and recovery time, and it seems necessary to take care to compensate if needed.

~The fact of developing a body(sensing organ, physical body) that is as flexible as the artificial brain, is also very important. **

:In robots, developing animal hardware that physical/sensing organs are fused with the current technical level, is a very big stumbling block.

:A converged approach is needed in the area of various sensing and control, which is now at the beginning level.

~It will take a lot of human 'effort, time, and various approaches', to create hardware made of 'body architecture, actuator, sensing organs' to be as highly efficient as the current higher organisms that have evolved functionally and environmentally over a long period of time. **

~If the above is hardware, the brain(learning, decision ability) and efficient access/utilization algorithms to memory will be the software. **

~Here is one more thing to consider, the brain part must be effectively evolved in each case in line with the evolution of the technology/performance of the physical area of sensing and actuator. It can be a presentation of methodologies for creating evolution by compressing the evolution of animals that have been very slow for billions of years into decades or even hundreds of years.

~Architecture = Cognition(sensing) + physical body + intuitive memory + memory access + 'batch & efficient & intuitive' learning ability

:The whole component of Architech must be developed evenly(without falling field) to create a meaningful humanoid robot.

2.3 Smart actuator (by muscle & motor)

~If robot can predict the strength required for the robot's posture and future movements, it will need to improve its time/energy efficiency in preparation for the proper tension/force adjustment of the muscles, and be prepared not to panic or lead to accidents from your own physical changes that are unpredictable. Let's apply similar smart algorithms on robots and smart machines/devices, like algorithms where animals use learning and memory.

~The most important thing to enable it is to recognize, perception, learning, experience, and memory. Recognizing the stairs and ascent in front of several meters in front of robot, preparing the power & operation necessary to pass through the road in advance, or learning & remembering how to turn various water valves or doorknobs, etc., present in different forms and structures for each location & environment, and then it will have to effectively utilize the accumulated learning memory of the past during its reuse time.

And when you remember each experience, you can expect a great effect on efficiency and accuracy if you remember it with the information of the location and environment.

:We sometimes don't notice the difference in height on stairs or paths, so we can't prepare for muscle strength or balance in our legs/body, so we can easily fall into dangerous situations. This is related to the attention to safety. Most animal organisms can respond to the most efficient and safe physical operation by accessing the learning memory in a variety of sensing and the environment with visual cognition in each environment in advance. In particular, aspects of his or her safety often operate as a very high priority.

~Efficient movement/behavior strategy based on convergence judgments such as movement/behavior goal, current posture, physical movement of each part(direction, speed, etc.), perception of the surrounding environment will be the basic. It seems that the use of basic knowledge algorithms for understanding each environment, and the efficient 'access, utilization, sharing, etc.' of learning/memories of movement/behavior in the environment, are necessary.

~When developing various behavior&decision algorithms of robots for the weight and contents prediction of the object to be carried and moved, door opening method, and optimal way to move, it will be basic to reflect factors such as efficiency, safety, urgency in the robot behavior, like a car that operates safely and efficiently through 'ADAS, self-driving sensing, transmission/accelerator/brake control'.

~Human absorption of a number of various learning information sharing, has a physical limitation in terms of time and storage capacity. However, compared to the obstacles that humans have, there are most cases where there are no such obstacles in the machine & device. In the world of smart and connected 'robots, machines, and devices', it is much easier to share the memories of such learning and experiences. I expect this to accelerate the development of human civilization by contributing once again to human civilization in terms of increasing humanity's time and improving efficiency.

The important thing here is that a large number of human beings must try to catch and disseminate the insights of the device & AI system to become individuals who evolve in harmony with the development of human civilization. **

~Ensuring durability considering the environment of each application

:Agricultural robots should be designed with sufficient consideration of the environment such as 'dust, water, outdoor, and night work'.

:Caregiver robot and elderly/chileren helper robots are designed to help people with maximum carefulness & security.

:Personal assistants and friend robots need flexibility and detailed sensing to share and analyze the user's experience in general life, so they need to act as adviser or companions. Sometimes you have to act as a bodyguard, so your quick movement and cautious response skills will also be considered.

2.4 Joints & Muscles & Flexibility (for shock absorption solution)

~Many devices, such as automobiles and machines, are using shock absorber such as spring or rubber for shock absorption purposes.

~Robots must also be properly designed with shock absorbing solutions that take into account the speed and various other activity characteristics such as 'amount of activity, activity area, use environments, etc.' of each application.

~Animals, including people, using the tension control of the muscles and bending of the joints in many activities to absorb most of the impact, and at the same time, utilize it for the control of the center of gravity. Recently, robots have already reached the stage of using joints and inertia and kinetic energy for its wonderful motion. I hope that the development of artificial muscles that will give them a lot of flexibility and efficiency, will be completed quickly.

~In each activity of many animals with joints, including people, constantly adjusting the bending angle of the joint and also adjusting the tension & power of the muscles, and finally using the flexibility of the bone connection. Such mechanisms are intended to minimize damage to the body physical part.

Even if developers don't have to learn physics, they can design robots that feel and respond to the pain of physical damage. It's easy. Therefore, to make it easier to develop robots, it seems that sensing techniques to feel the pain of 'muscles, joints, etc.' and corresponding algorithms are needed.

~Animals seem to use two algorithms in a fusion for physical behavior. One is an algorithm for applying the prior learning results, such as 'experience, learning, cognitive, memory, etc.' of each organism for the efficient control of the joints and muscles, the second is an algorithm that senses pain & margin in real time and corresponding algorithms from the physical part, such as the control of the muscles or joints. This animal fusion approach would also be good for the development of robots. I think there is probably no dissent to this idea. Hehe.
**

~Above all, it seems that a chemical-based approach should be a priority for the understanding of the 'structure, growth, operation' of the animal muscles. The fact that the chemical approach to the field of robotics where physics is heavily applied is only recently started, it is the cause of the fact that a good muscle actuator has not yet been developed. Just as convergence is a trend in ICT, it seems that convergence collaboration in basic sciences is urgently needed in the field of 'physics, chemistry, biology, etc.'

~The evolution from a simple voice service that does not require action to a high-level behavioral service will be compared to the increase in value that adds visual sensing to auditory sensing. This is because the AI voice service alone cannot serve as a bodyguard, a caregiver, or a friend who walks on behalf of a pet dog or walks a pet dog. To act together in a free human society, robot will need to have flexible HW and animal motion functions. *

2.5 Fish fins in Water vs. Bird Wings in Air

~The organ that makes the propulsion and posture control of the fish in the underwater environment will be a combination of fish fins and tails instead of propellers.

~It seems to be able to utilize the propulsion and posture control of the fish robot using the tail-shaped main propulsion parts and fins of the flexible material, such as rowing a non-motorized boat instead of a small propeller that rotates rapidly.

~In comparison of the components of the space environment in the atmosphere & water, the density of the water is said to be approximately 800 times the level of the air. Therefore, it is said that a bullet that goes between 800 and 2,400 meters in the air, can only travel from 1 to 3 meters in the water. In those environments of different densities, the basic wisdom of physics will be required for devices or life creatures with volume and weight to move. Like the observation of bullets, it is relatively efficient to obtain the propulsion and lift force at a slower speed movement in dense water, but it's inefficient when moving at high speeds. And, due to the small density difference between heavy water and fish/device, the energy loss by gravity is basically very small when moving in the water.

On the other hand, in the low density atmosphere, a lot of energy is consumed in the area that creates a lift force against gravity because the density difference between the atmosphere and the bird/device is very large. So, in order to make that point efficient in the atmosphere, birds and planes use wings and high speeds to make it easier to lift against gravity. And since using a low density of air, propulsion is obtained usually through a high-speed rotating propeller or jet engine at less torque than the propulsion system in the water.

~If I lose a lot of weight, strengthen my arm muscles, and then flapping my arm with a pretty big flexible wings, I think I'll be able to fly like a bird. However, in order to reduce the weight like birds, I have to be transformed into a light bone with osteoporosis, remove heavy teeth and jawbone, and the food and feces in my body should be minimal, and strengthen the arm to mount the wings and the muscle associated with it, and the weight and size of the legs will need to be reduced to less than 30%. The imagination of my transformation into a bird makes me laugh. Hehe.

~Using the arm to simulate a propulsion of a flexible fish tail shape, it seems to be able to get maximum propulsive force, immediately after the left/right (or up/down) direction changes to the opposite, when the palm part, after the wrist, is flexible. This is why flexible materials should be used when using the body as a propulsion parts, like fish. It seems that a flexible physical organ is needed, including 'proper tension, sensing to feel the pressure or frictional force of air/water, and fine control functions'. **

~A penguin that is good at flying in the water, gains the main momentum with its wings. Fat penguins can fly in the water, because they require little energy to fight gravity, like other sea creatures. And, in order to fly in low-density air, the birds need large wings, but in the case of penguins that wing in dense water, their small wings(arms) are sufficient to fly(swim).

2.6 Thinking about the design of physical organs for fish/bird robots

~The tension factor is most important in the design of flexible propulsion parts.

~There is a limitation in efficiency with a fixed mechanical structure joint that has been popularly used by mankind in the connection between the main body and the flexible propulsion parts.

~Like animals, the flexible propulsion part is connected to the main skeleton with the structure of cartilage & ligament & muscle, for maximum efficiency.

~Flexible propulsion part of fish fin type for control & low power, is connected to the strong muscles of the skin, without direct connection to the skeleton

:In swimming(especially butterfly), the flexible movement of the arms and hands seems to indicate its importance.

~The experience for realizing tension

:It was easy to see the importance of tension when using your arms, legs or head while enjoying sports. It is very important in motion that requires thrust such as smashing, kicking, swimming, gymnastics, skiing, bird-wings, etc. Most sports masters have already realized that point.

:In the YouTube video "Maksim Mrvica - Claudine original", I think you can get a hint from the ballerina's beautiful arm movements.

~The use of such efficient tension is expected to be the key material/part of the future that makes the physical organs of animal robots/devices. **

2.7 The validity of an easy architecture approach that mimics the simplicity of a single animal

~It may take a very long time to make bio-chips or bio-brains. The birth of artificial organisms that is impossible/difficult to control, or not-difficult/simple structure, will be witnessed not too difficult by bio & genetic engineering. It will also be an important task to easily analyze the level of 'long-term reliability, control reliability, energy supply, learning, etc.' associated with these artificial organisms. Of course, like the history of computer development, I think that even if the beginning is weak, it may develop into a higher animal level, in 20-50 years.

~Certainly, with the prior development of such biochips, the completion of detailed programming for everything that the brain can do, such as 'interworking of 'learning, sensing, physical movement', memory, ethics, greed, fear, love, friendship, joy, anger, sorrow, and pleasure', that is, the emotions of humans or animals, will be an important base(asset) just before connecting them very closely and creating a comprehensive solution for efficient operation. **

~Unexpectedly, like simple animals/people(?), let's approach with an algorithm that simplifies 'space, events, and thoughts' around things that are very important to each creature. **

~Robots with capabilities such as billions of people's wisdom or Superman-like, are so far away that we aim at small wisdom or small sensing/physical capabilities that mimic the level of an average person or one puppy/cat, and let's also approach with a methodology that invests as much as the learning time of each real animal. *

1. AI(ARTIFICIAL INTELLIGENCE) vs. ROBOT

1.1 The big difference between artificial intelligence and robots **

~As an independent object's qualification(self), adding elements such as physical force and behavior to AI(sensing, decision, memory, learning) is defined by the concept of a robot.

~When robots and devices are added to AI with 'physical power, flexible behavior, various sensing', it will provide blue ocean-class value that is close to the new world that has never existed in human civilization society. "Care services for 'children, patients, the elderly', cleaning, work, gardening, washing dishes, cooking, etc." could be replaced with non-discontent devices devices, and will be able to play the role of 'personal assistant, friend, companion, professional'.

~If you have enjoyed swimming or bowling freely, I will soon be able to enjoy ping-pong and tennis with your own robot friends (at the time/place you want to) in the future. A machine for ping-pong has already been developed. If you apply and learn about robot sensing, posture control and physics more precisely, you will have robots playing ping-pong soon.

~However, we will have to be prepared for the use of such robots by criminals or bad people. As with the advent of cars and the Internet, there may be a lot of vague fears, but the value of a robot's contribution to the human society with such capabilities will be as much as the difference in the value of a wagon and a car.

~We must learn from history that a strategy of daring to use such good things and suppressing bad use has always led to the development of human civilization.

~Like creatures such as humanity, robots must operate in the form of stand-alone AI, not centralized Cloud AI, in order to qualify for each independent object.

~From the robot's point of view, using centralized Cloud AI will be very dangerous or unstable for its sustainability or growth into an independent ego. However, for humanity wants to safely utilize the robot, it is expected to act as a safety plate in the utilization of robots, it also acts as a feature that makes human society easily applicable to their society.

~Compared to efficient organisms, because there is a limit such as 'physical organ, energy, AI(sensing, decision, memory, learning)' in the configuration and operation of the robot, a simple stand-alone part will perform a limited performance and role, advanced or dangerous areas are expected to be utilized under a lot of human scrutiny.

1.2 Can artificial intelligence create demons? **

~The Ministry of Security(global safety, human rights and peace-related organizations) should have access and administrator authority over AI.

~AI itself can be a major threat if it is programmed to maintain its own persistence, control its energy supply, and even hack or create&utilize robots or devices, and turn off AI to disrupt or attack human social activities. Of course, I think it will take a little more time to build such an environment.

~Such a dangerous program is, after all, made by a demonic bad guy. If a small number of demons complete a program with a perfect survival instinct and greed and destructive basic properties that cause enormous repercussions, and even create and utilize the minimum infrastructure for it, human society may be at great risk. In this case, like the sci-fi film, the next 3rd World War could be a war between the Allies of humanity and robots.

~For a bad developer with such skills or an bad employer employing such a developer to come out, at least the ability to complete a perfect program after completing a bioprocessor, which includes functions such as efficient 'sensing, decision, memory, learning', and building an environment through investment in many areas.

And even if he or she is a genius like Einstein, I think it is very unlikely that a bad small group will make a multi-faceted convergence program without making a mistake. It would be as unlikely as the existence of extraterrestrial life, such as the current high level of human civilization.

~When a result is made with a great number of powers, it makes us to alleviate fear that many mature wisdom, ethics, culture, and right values are never in favor of demonic making.

1.3 Robot's required specifications

~The physical On/Off switch of the energy source will need to be mounted on the outside like the emergency button.

~When possessing a certain level of 'power(strength, output), weapons, weight, behavioral capacity, speed, etc.', it must be identified.

:GPS required for tracking function(like a criminal ankle tracker).

:In case of emergency, risk removal function by remote control is required.

~Software

:Screening and licensing processes are needed to ensure that there is no bad intent.

:The difference between a fixed robot and a moving robot must exist.

:Anti-hacking algorithms must be built-in.

:Physical replacement of memory containing the program or online firmware update requires strict permission and management.

:Otherwise, there is a possibility that a robot that behaves badly, such as Microsoft's abusive AI, will be born.

:Proper learning is so important.

:Be prepared to operate only with basic programs that do not have smart social/ethical learning functions.

1.4 The difference between people and robots

~Difficulty level of problem to solve, feels different

:In the 'memory, search, comparison' ability of Big Data and the determination of digital matching scores, robots are excellent

:Simultaneous/intuitive 'approach, decision, predictive' power of memory data in various fields, creative convergence thinking skills based on converged memory in various fields, human and social decision-making skills, etc. are excellent for human beings.

~Social infrastructure(architecture, tools, transportation, etc.) created by humanity are absolutely unfavorable to the robot.

:I think it's time to consider the social infrastructure for robots.

:Because robots are difficult to use this social infrastructure for humanity, I believe that robots can not threaten humanity for the time being, even if the robot's other limitations will be eliminated in the near future.

0. PROLOGUE

~In the natural environment of animals or human artificial social environments that have been operating with muscles and bio-brains for hundreds of millions of years, there will be large and small limitations for devices such as robots controlled by digital motors to play many roles flexibly. I am sharing some approach ideas to pioneers who feel that limitation, and I expect them to make additional contributions to the future of humanity.

And, it mainly deals with convergence technologies such as sensing organs, brains, and flexible bodies that mimic animals from a macro perspective.

~"Personal leisure time" is being increased by the development of human civilization such as 'mechanization, automation, industrialization, AI', and "the time of old age" is also being greatly increased due to the prolonged lifespan due to medical development. The moments of a lonely or needy will be more likely to increase in our life's leisure time. A robot that includes all of the specs like "flexible physical body, biological brain based computing and memory, efficient convergence sensing" that resembles an animal, that will become a "friend or personal assistant" for those moments/times of high value. That's also why it looks attractive.

The biological brain part will be the bottleneck. However, I think that part is worth approaching as a two-sided strategy of the development of SW architectures at various levels and mid- and long-term investment in understanding of bio-brain. It will evolve in the direction of installing appropriate services and software for each application field in HW architecture of various performances. I believe that this contents are also offering a variety of approaches from a business and developer perspective for devices and services that will support humanity in responding to these changes in the environment.

~When studying, discussing, and imagining, this is a collection of memos taken before the flashy ideas of the moment evaporate into the air. I hope that users of this content understand the imperfection, considering that this content is focused on efficiently sharing individual ideas without going through a professional publishing system.

~If you just remember deeply in your brain with only one

sentence saying that the size of a personalized robot market, such as a personal assistant or a friend robot, may exceed the automotive market, and then you have already mined the $100 value from this content.

~Imagine 'products and services and the future' from the user/consumer's perspective and try it. If you create something(product, service, strategy, etc.) for the 'users, consumers, people, members' that make up the essence of the market and society, you will be much less likely to fail. Imagine and prepare the devices/services and the features in it that you and the average person want to have.

~If this content is very helpful and you get a jackpot, I hope you will return 30% of your profits to the field of social contribution based on your philosophy and values.

~I am grateful for 'the experts in each field, the people around me, and the environment' who helped bring out such ideas by sharing their numerous wisdoms and experiences.

~The mark '* ~ *****' in the contents is an expression of the level of feeling I want to emphasize when I make a note.

~Finally, I am grateful to the machine translation service platform companies (Google, Microsoft, Naver Papago) and the developers.

:Machine translation weight (40% in English)

ABOUT THE AUTHOR

~Since I was a child, I have enjoyed applying electrical appliances to create things that are fun and beneficial to my life, and my dream was to become a scientist. Also, rather than keeping the evil laws created by society, I'm living with the philosophy that evil laws must be corrected in the right direction. In addition, it seems to be expanding the area of interest and maintaining a curious state of interest all the time. At this rate, I am even afraid that if I study for another 50 years, I will become a god. Hehe~

~There is a big market where most of the human members are users, a device that their users always carry and use, and a black hole for all the devices and hub device in each individual life, that is a mobile phone or smartphone. I seem to have enjoyed and worked for 18 years in cell phone & smartphone R&D. Among them, I have experienced HW R&D for 13 years, and for 5 years, has been in charge of various areas that hope to be useful on smartphones such as 'components such as sensors, technologies, SW solutions, AP(application processor), new technologies, and technology strategies'. It seems to have expanded the diversity in earnest by collaborating with experts in the field of numerous components and various solutions.

~After coming out of the company due to changes in the market environment of smartphone devices, I think that I was able to acquire the different talent for imagination that enables convergence applications, while studying "AI, autonomous vehicles, drones, blockchain, startups, global economic issues, environment, energy, peace and fairness" through 'seminars, conferences, forums, etc.' with experts, for six years, as a personal qualification. I would like to share the idea of a convergent imagination that may be lacking for a large number of people who lack time in a specialized/divided areas.

~The flashing ideas of every moment based on such imagination have been noted in a smartphone or note in my hand. Also, it seems to be a good source to experience various human life environments. I hope that the success of this content business will soon lead to the life of the rich man who runs Noblesse Oblige.

~Finally, the author has a philosophy that each beneficial wisdom and philosophy should be widely disseminated and persuaded in a majority of human society. It is because I believe that many people can obtain wisdom, culture, and values that can continue to develop human civilization and go the right way through such a process. I aim to increase humanity's most precious free time, reduce conflict by providing fairness, and provide opportunities for something for everyone, as human civilization, culture, and values in various fields mature. I also share my content to create the devices and services I desperately want to use and the human society I dream of.

www.ingramcontent.com/pod-product-compliance
Lightning Source LLC
Chambersburg PA
CBHW071409210526
45465CB00001B/313